CURTIS'S
BOTANICAL MAGAZINE,

COMPRISING THE

Plants of the Royal Gardens of Kew

AND

OF OTHER BOTANICAL ESTABLISHMENTS IN GREAT BRITAIN;
WITH SUITABLE DESCRIPTIONS;

BY

JOSEPH DALTON HOOKER, M.D., C.B., P.R.S., F.L.S., &c.

D.C.L. OXON., LL.D. CANTAB., CORRESPONDENT OF THE INSTITUTE OF FRANCE.

VOL. XXX.
OF THE THIRD SERIES;
(Or Vol. C. of the Whole Work.)

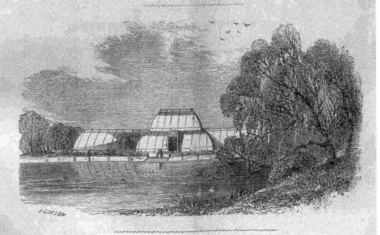

"Moreover, in these fayre offspringes of Nature there is a charm removed from their beautie, the which appealeth to the harte; many find in them the emblems of some excellent qualitie or tender feelinge."—OLD HERBAL.

LONDON:
L. REEVE & CO., 5, HENRIETTA STREET, COVENT GARDEN.
1874.
[All Rights reserved.]

LONDON:
SAVILL, EDWARDS AND CO., PRINTERS, CHANDOS STREET,
COVENT GARDEN.

TO

GEORGE MAW, ESQ., F.L.S., F.G.S., &c.
OF BENTHALL HALL, BROSELEY.

My dear Maw,

Allow me to dedicate to you this volume of the "BOTANICAL MAGAZINE," as a tribute to the value of your exertions in introducing hardy herbaceous plants into English gardens. No one of late years, or perhaps ever, has collected with his own hands so many of these for transmission to England, cultivated them with more success, or distributed them with more liberality—as the pages of this work to some small extent testify.

Allow me also in this dedication to refer to that delightful excursion to the coasts of Marocco and the Greater Atlas, that we made in company with our friend Mr. Ball, and which resulted in the introduction of so many interesting plants not hitherto known to English gardens.

Believe me,

Most sincerely yours,

JOS. D. HOOKER.

Royal Gardens, Kew,
Dec. 1, 1874.

Third Series.

No. 349.

VOL. XXX. JANUARY.   [Price 3s. 6d. col^d. 2s. 6d. plain.

OR No. 1043 OF THE ENTIRE WORK.

CURTIS'S
BOTANICAL MAGAZINE,

COMPRISING

THE PLANTS OF THE ROYAL GARDENS OF KEW,

AND OF OTHER BOTANICAL ESTABLISHMENTS IN GREAT BRITAIN,
WITH SUITABLE DESCRIPTIONS;

BY

JOSEPH DALTON HOOKER, M.D., C.B., F.R.S., L.S., &c.

Director of the Royal Botanic Gardens of Kew.

Nature and Art to adorn the page combine,
And flowers exotic grace our northern clime.

LONDON:
L. REEVE & CO., 5, HENRIETTA STREET, COVENT GARDEN.
1874.

[All Rights reserved.]

Mo. Bot. Garden,
1897.

Just published, price 7s. 6d.

A SYNOPSIS OF THE BRITISH MOSSES,

Containing Descriptions of all the Genera and Species (with localities of the rarer ones) found in Great Britain and Ireland. Based upon Wilson's Bryologia Britannica, Schimper's Synopsis, &c. By CHARLES P. HOBKIRK, President of the Huddersfield Naturalist's Society.

"A singularly handy book. . . . To the ordinary student who has made some progress, it has some advantages even over Wilson's Bryologia."—*Scottish Naturalist.*

L. REEVE & Co., 5, Henrietta Street, Covent Garden.

BOTANICAL PLATES;
OR,
PLANT PORTRAITS.

IN GREAT VARIETY, BEAUTIFULLY COLOURED, 6d. and 1s. EACH.

List of nearly 2000, one stamp.

L. REEVE & Co., 5, Henrietta Street, Covent Garden.

FLORAL PLATES,
BEAUTIFULLY COLOURED BY HAND, 6d. EACH.

A New List of 500 Varieties, one stamp.

L. REEVE & Co., 5, Henrietta Street, Covent Garden.

NOW READY.

BENTHAM AND HOOKER'S

GENERA PLANTARUM.

Part IV, being the first part of Vol. II., comprising Caprifoliaceæ to Compositæ. Price 24s.

L. REEVE & Co., 5, Henrietta Street, Covent Garden.

"THE GARDEN,"

A Weekly Illustrated Journal devoted solely to Horticulture in all its branches.

THE GARDEN is conducted by WILLIAM ROBINSON, F.L.S., Author of "Hardy Flowers," "Alpine Flowers for English Gardens," "The Parks, Promenades, and Gardens of Paris," &c., and the best Writers in every department of Gardening are contributors to its pages.

The following are some of the subjects regularly treated of in its pages:—

The Flower Garden.	Hardy Flowers.
Landscape Gardening.	Town Gardens.
The Fruit Garden.	The Conservatory.
Garden Structures.	Public Gardens.
Room and Window Gardens.	The Greenhouse and
Notes and Questions.	The Household.
Market Gardening.	The Wild Garden.
Trees and Shrubs.	The Kitchen Garden.

THE GARDEN may be obtained through all Newsagents and at the Railway Bookstalls, at 4d. per copy. It may also be had direct from the Office at 5s. for a Quarter, 9s. 9d. for a Half-year, and 19s. 6d. for a Year, payable in advance; and in Monthly Parts. Specimen Copies (post free) 4½d.

37, Southampton Street, Covent Garden, W.C.

BOTANICAL MAGAZINE.

THIRD SERIES.

NOTICE OF RE-ISSUE.

SOME portions of the above work being out of print, and complete sets very difficult to obtain, the Publishers have determined to reprint so much as will enable them to complete a few copies as they may be subscribed for; and to meet the convenience of Subscribers, to whom the outlay at one time of so large a sum as a complete set now costs is an impediment to its purchase, they will commence a re-issue in Monthly Volumes, thus spreading the cost over a period of two years and a half. The price of the volumes will be 42s. each as heretofore, but to Subscribers for the entire series, 36s. each. The first volume will be ready January 1st, 1874.

As the reprinting will be carried no further than is required to meet the actual demand, early application is necessary to prevent disappointment. Persons preferring to take a set complete at once may do so at the subscription price of 36s. per volume. Of the THIRD SERIES, twenty-eight volumes are completed; the twenty-ninth will be ready in December.

The BOTANICAL MAGAZINE, commenced in 1787 and continued with uninterrupted regularity to the present time, forms the most extensive and authentic repertory of Plant

History and Portraiture extant. The THIRD SERIES, by far the most valuable, comprising all the important additions of the last thirty years, contains about 2000 Coloured Plates, all from the characteristic pencil of Mr. W. Fitch, of Kew, with Descriptions, structural and historical, by Sir William and Dr. Hooker.

LONDON:
L. REEVE & CO., 5, HENRIETTA STREET, COVENT GARDEN.

FORM FOR SUBSCRIBERS.

To Messrs. L. Reeve & Co., Publishers,
    5, Henrietta Street,
        Covent Garden.

Please send to the undersigned the BOTANICAL MAGAZINE, THIRD SERIES, *in Monthly Volumes, or complete,* at 36s. per Volume.*

*Name*_____

*Address*_____

_____

*Date*_____

*Conveyance*_____

* Subscribers will be good enough to indicate in which form they desire to receive the work, by striking out the words indicating the other form.

TAB. 6074.

SAXIFRAGA PELTATA.

Native of California.

Nat. Ord. SAXIFRAGACEÆ.—Tribe SAXIFRAGEÆ.

Genus SAXIFRAGA, *Linn.*; (*Benth. & Hook. f. Gen. Pl.*, vol. i. p. 635).

SAXIFRAGA *peltata;* rhizomate crasso repente, foliis omnibus radicalibus longe petiolatis erectis amplis peltatis orbiculatis ambitu 6–10-lobatis lobis inciso-dentatis, scapis nudis elongatis, cymis corymboso-capitatis multifloris glandulosis, calycis tubo brevissimo obconico-campanulato, sepalis oblongis obtusis, petalis ellipticis obtusis, filamentis subulatis, carpellis liberis in stylos breves attenuatis, stigmatibus dilatatis.

SAXIFRAGA peltata, *Torr. Mss. in Benth. Plant. Hartweg,* p. 311; e in *Bot. U. States Explor. Exped.* Pl. v. (*without text*); *Engler Monog. Gatt. Saxifr.* 108 *Walp. Ann.,* vol. vii. p. 891.

One of the largest species of the genus, and a very curious one, though far from being as handsome as many others. Variable as the foliage of the Saxifrages is, the present is the only one known in which that organ is completely peltate, and like many other peltate-leaved marsh and water-loving plants, this is stated to be found on the margins of streams and in the water itself. I have seen indigenous specimens gathered in the Sacramento Mountains by Hartweg, who discovered the species; in the Mendreino county, California, by Prof. Bolander, of San Francisco; and others collected by Lobb without a locality. A very singular form, either a variety or different species, is in the Hookerian Herbarium from Clear Creek in North California; it has the almost glabrous cyme broken up into a distantly branched panicle, the branches of which have short rounded bracts at the base, and has inflated much rounded carpels abruptly terminated with short styles; the fruit figured in the "Botany" in the United States Exploring Expedition resembles this, and not the narrower attenuated fruit of Hartweg's, Lobb's, and the cultivated plants. Engler in his Monograph of *Saxifraga*

JANUARY 1ST, 1874.

makes a section of it (*Peltophyllum*) founded on the shape of the leaf, and on the carpels opening above only, but the latter is an error, for the carpels open to the base both in this plant and in that figured by Torrey.

I am indebted to Messrs. Downie, Laird, & Laing for a specimen of this fine plant, which flowered with them in April 1873; it would probably attain a greater size if planted in or near water and become a very conspicuous and attractive object.

DESCR. *Rootstock* as thick as the thumb, creeping, partly buried in the soil; green with large broad leaf-scars; the tip clothed with the broad green stipular leaf-sheaths, which are rounded with membranous pink-margins. *Leaves* all subterminal, erect; petiole one to two feet long, cylindric, as thick as a goose-quill, glandular-pubescent; blade orbicular, peltate, six inches in diameter, 6–10-lobed, the lobes cut and sharply-toothed, upper surface dark green with a deep funnel-shaped depression at the centre where attached to the petiole, pale beneath. *Scape* equalling or exceeding the leaves, terete and glandular like the petiole. *Cyme* three to five inches in diameter; subcapitate, repeatedly branched, ebracteolate, glandular-pubescent. *Flowers* one half inch in diameter. *Calyx-tube* very short, between obconic and campanulate; lobes 5, reflexed, oblong, tip rounded. *Petals* longer than the sepals, elliptic, rounded at both ends, white or very pale pink. *Stamens* equalling or exceeding the petals, filaments broad subulate, anthers small broad. *Carpels* two, nearly free, narrowed into short stout styles; stigma dilated. *Fruit-carpels* one-third inch long, narrowed into the style. *Seeds* large, subcuneate, angled, compressed, brown.—*J. D. H.*

Fig. 1, Flower, with petals removed :—*magnified.*

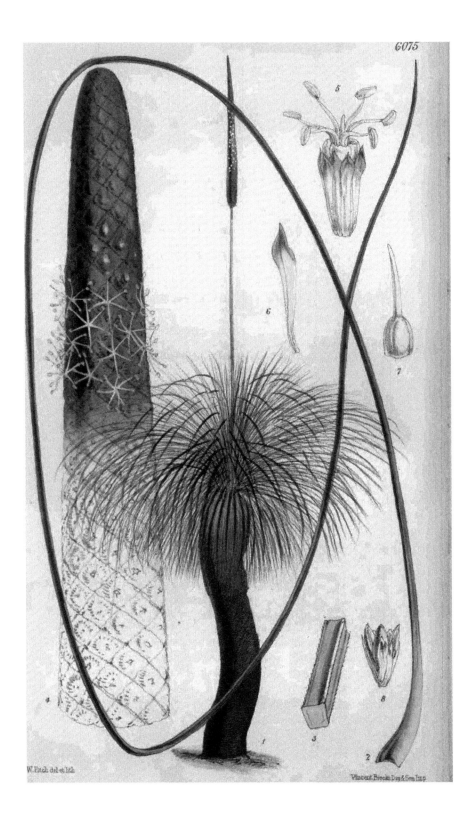

TAB. 6075.

XANTHORRHŒA QUADRANGULATA.

Native of South Australia.

Nat. Ord. JUNCEÆ.—Tribe XEROTIDEÆ.

Genus XANTHORRHŒA, *Smith;* (*Endl. Gen. Plant.,* p. 152).

XANTHORRHŒA *quadrangulata;* trunco arboreo, foliis gracilibus e basi paulo dilatato glaberrimo filiformibus rectangule tetraquetris glaucis angulis scaberulis, scapo 2-6 pedali, spica 3-4 pedali, bracteis numerosis angustis apice rhombeo-dilatatis acuminatis, sepalis albis paulo longioribus æqualibus anguste spathulatis obtusis v. cuspidatis valvatis; petalis, lineari-oblongis obtusis cuspidatis, staminibus longe exsertis divaricatis, capsula perianthio longiore.

XANTHORRHŒA quadrangulata, *F. Muell. Fragment. Plant. Austral.,* vol. iv. p. 111.

The Grass-gum trees are amongst the most remarkable vegetable features of that country of wonderful vegetable forms, Australia; and it is with great satisfaction that we now figure in the *Botanic Magazine* a second species of a genus so rare in cultivation. It is a native of South Australia, where it inhabits rocky hill-ranges, and was sent to Kew by Dr. Schomburgk, the energetic Director of the Adelaide Botanic Garden. Shortly after its arrival, the trunk which is four feet high, slowly developed its fresh green leaves, which steadily increased in number and length till the plant had the appearance given in the plate; the flower-stem and Typha-like spike commenced to emerge about July of last year, and attained its full development in September, when the flowers began to expand from below upwards, and a full month elapsed before all had opened. During flowering time a copious honey-like secretion was exuded, which hung in great tear-like drops to the brown spike. A few of the ovaries have swollen, and indeed matured, but the seeds have not been fully formed.

Dr. Engelheart, of Gawlor-town, South Australia, an ardent Horticulturist, informs me that the *Xanthorrhœas,* of which there are two species in that district (*X. semiplana* and *quad-*

rangulata), like a rich fern-soil, mixed with a good deal of fine black sand, and drive their straggling roots into crevices of rocks 20–30 feet down amongst the accumulated vegetable soil. Young plants have a very pretty appearance, resembling a *Gynerium*, but growing older, and periodically subjected to bush fires, all the leaves but the central are consumed, and an ugly charred and blackened stump with a tuft of leaves remains.

About fifteen species of *Xanthorrhœa* have been discovered, of which the *X. Hastile* of New South Wales (Tab. nost. 4722), is the best known, from the uses of its long peduncles, which attain twenty feet in height, as spear-shafts, and for the rich red-brown astringent resin which forms between the densely compacted bases of the leaves, and which has been used as a substitute for gum-kino. It is often called the Blackboy, and a native boy with a tuft of grass on his head placed amongst a group of them, is, from a little distance, with difficulty distinguished from the surrounding trunks. Another species, *X. pecoris*, F. Muell., of West Australia, forms a staple fodder for cattle during a good part of the year. Several species are cultivated at Kew—viz., *X. quadrangulata, semiplana*, F. Muell., *Hastile*, Br. and *minor*, Br., with others not yet in a sufficiently advanced condition for determination.— *J. D. H.*

Fig. 1, Whole plant:—*reduced in size;* 2, leaf:— of the *natural size;* 3, transverse section of leaf:—*magnified;* 4, upper part of spike:—of the *natural size;* 5, flower and bracts; 6, bracts; 7, ovary; 8, fruit:—*all magnified.*

TAB. 6076.

STEUDNERA COLOCASIÆFOLIA.

Native of South America.

Nat. Ord. AROIDEÆ.—Tribe ASTEROSTIGMEÆ.

Genus STEUDNERA; (*Koch in Regel Gartenflora*, 1869, p. 323).

STEUDNERA *colocasiæfolia;* caudice brevi crasso vaginis brunneis tecto, foliis longe petiolatis, petiolo tereti, lamina peltata concava ovato-oblonga acuminata basi modice emarginato-2-loba, subtus glauo-viridi maculis brunneis latis infra nervos irrorata, pedunculis petiolo brevioribus et tenuioribus viridibus, spatha ampla tota aperta late ovata acuta recurva flava disco pallide et sordide rufo-purpurea, spadice spatha multo breviore obtuso parte fœminea dorso spathæ fere toto adnata, ovariis confertis staminodiis brevibus clavatis circundatis hemisphericis 2-locularibus, stigmate sessile discoideo 5-gono, antheris columnæformibus late truncatis loculis 7–8 parallelis connectivo columnari carnoso longitudinaliter adnatis.

STEUDNERA colocasiæfolia, *Koch*, l. c.

According to Regel and Koch this singular Aroid is a native of South America, whence it was imported by Linden, if, as I venture to think, the plant here figured and for which I an indebted to Mr. Bull, is the *Steudnera colocasiæfolia* of Koch. Of this a comparison with Regel's description and plate would leave no doubt in my mind, were it not that Mr. Bull's plant has many staminodes and a 2-celled ovary, whilst Koch's has but one or two staminodes, and a 5-celled ovary; the number of staminodes is very likely to be variable, as is frequently the case with arrested organs, and our plants having, like Koch's, 5 rays to the stigma would indicate the probability of there being sometimes as many cells to the ovary.

I regret having no information as to the exact habitat of this plant. Mr. Bull believes that he received his specimen from Calcutta, but it is certainly not an Indian form. It belongs to Schott's section or tribe of *Asterostigmeæ*, and its allies are for the most part American; it, however, closely resembles in the form and colouring of the foliage a very

JANUARY 1ST, 1874.

ornamental Aroid (*Colocasia Jenningsii*), which I found in the Khasia mountains. The specimen here figured flowered in Mr. Bull's nursery in May, 1873.

DESCR. *Rootstock* one to two inches high, and one and a half in diameter, clothed with brown sheaths. *Leaves* few, six to ten inches long, terminal, oblong-ovate, acuminate, peltate, concave, with a shallow notch at the base, midrib strong, as are two nerves that proceed backwards from the insertion of the petiole; lateral nerves numerous, spreading; upper surface dark green, under glaucous green, with dark brown blotches between the nerves; petiole one foot long terete, green. *Peduncle*, shorter and more slender than the petiole, terete. *Spathe* four inches long, broadly-ovate, acuminate, quite open, slightly concave, recurved after opening, yellow, with a suffused pale red-purple disk. *Spadix* one and a half inches long, upper one-third free, subclavate, obtuse, clothed densely with hexagonal anthers; lower two-thirds dorsally adnate to the spathe, densely clothed with ovaries, and each surrounded by five to eight short clavate staminodes. *Anthers* shortly columnar, apex of connective flat dilated; cells seven to eight, linear, parallel, surrounding the thick connective. *Ovary* hemispheric, 2-celled; stigma discoid, 5-angled; cells several-ovuled, ovules attached to the axis.—*J. D. H.*

Fig. 1, Spadix and its attachment:—*of the natural size*; 2, anther; 3, ovary and staminodes; 4, transverse section of ovary:—*all magnified*.

Tab. 6077.

MESEMBRYANTHEMUM truncatellum.

Native of South Africa.

Nat. Ord. Ficoideæ.—Tribe Mesembryeæ.

Genus Mesembryanthemum, *Linn.*; *(Benth. & Hook. f. Gen. Plant.*, vol. i. p. 853).

Mesembryanthemum *truncatellum;* obconicum, 1-3 poll. diametro, acaule, glaberrimum, glaucum, crassum, pallide viride, foliis ad 4 decussatim oppositis latissime cuneatis basi connatis appressis apicibus dilatatis, latissime truncatis, vertice lunatis convexis pallide bruneis depresso-tuberculatis colore saturatiore irroratis, basi vestigiis membranaceis fugaceis foliorum vetustorum vaginatis, floribus majusculis 1½ poll. diametro solitariis sessilibus, ovario inter folia 2 summa compresso, calyce 5-6-fido segmentis obtusis, petalis 2-serialibus numerosissimis anguste linearibus stramineis, staminibus perplurimis segmentis calycinis æquilongis, stylis ad 5 gracilibus, apicibus incurvis.

Mesembryanthemum truncatellum, *Haworth Miscell. Nat.*, p. 22; *Ait. Hort. Kew, Ed. 2*, vol. iii. p. 213; *Haw. Synops. Plant. Succ.*, 203; *DC. Prodr.*, vol. iii. p. 417.; *Harv. & Sond. Fl. Cap.*, vol. ii. p. 392.

Though differing in some respects from the published description of *Mes. truncatellum,* I have little hesitation in referring the vegetable oddity here figured to that plant. Thus Harvey, on what authority is not stated, describes it as only half an inch in size, whereas Haworth (the author of the species) calls it the "great dotted Dumplin," which implies that it is the largest of its allies, of which the smallest, *M. minutum,* is fully half an inch in size. Again, Haworth, in his original description (Misc. Nat.), describes the ovary as never extruded beyond the surface of the plant, as in our specimen; but in his Synopsis he describes it as exserted, and in this he is followed by De Candolle and Harvey. Lastly, the calyx is said to be 5-fid in the original description, and in De Candolle's Prodromus, and Harvey's Flora; but 4-fid in the Hortus Kewensis: in our plant it is 6-fid.

January 1st, 1874.

M. truncatellum was introduced into Kew in 1795, by F. Masson, one of the travellers attached to the establishment, and is described as flowering in November; our plant was sent by Principal MacOwan, of Gill College, Somerset East, and flowered in October 1873; it is a very rare species, was unknown to the Prince Salm Dyck, and is hence not included in his magnificent work; it has never before been figured.

DESCR. *Plant* forming tufts of pale glaucous green, obconic, truncate, translucent fleshy masses, one to three inches in diameter, with a flat or convex rather tubercled brown surface; each branch on plant, consisting of four leaves in opposite pairs, placed cross-wise. *Leaves* very fleshy, broadly cuneate, connate to above the middle; back convex; face convex; crown lunate, brown, mottled, convex; the first formed fleshy leaves, after developing another pair between them, shrink into a mere membranous pellicle that sheaths the base of the younger pair. *Flower* solitary, sessile, one and a half inches in diameter. *Calyx-tube* sunk and tightly-wedged between the two uppermost leaves; limb 5–6-cleft, lobes obtuse, tinged purple. *Petals* in two series, very numerous, straw-coloured. *Stamens* very numerous, anthers yellow. *Styles* 5, slender, tips uncurved.—*J. D. H.*

Fig. 1, Leaf; 2, flower with two sepals, and petals of same side removed:—*both magnified*.

TAB. 6078.

COLCHICUM SPECIOSUM.

Native of the Caucasus.

Nat. Ord. MELANTHACEÆ.—Tribe COLCHICEÆ.

Genus COLCHICUM, *Tourn.*; (*Endl. Gen. Plant.*, p. 137).

COLCHICUM *speciosum*; robustum, hysteranthum, cormo magnit. judglandis, foliis 4-5 late elliptico-lanceolatis, perianthii late purpurei tubo crassitie pennæ anserinæ 6-12-pollicari, limbo 5 poll. diam. segmentis ellipticis apice rotundatis concoloribus non tessellatis, antheris elongatis flavis, stigmatibus subunilateralibus integris apicibus incurvis perianthii segmentis multo brevioribus.

COLCHICUM speciosum, *Stev. in Mem. Soc. Nat. Mosc.*, vol. vii. p. 265, t. 15; *Kunth Enum.*, vol. iv. p. 139; *Hohen. Enum. Pl. Talusch.*, p. 23; *Koch in Linnæa*, vol. xxii. p. 258; *Ledeb. Fl. Ross.*, vol. iv. p. 204.

The largest known species of the genus, and a very handsome one, a native of the countries bordering the Caucasus range on the south, and extending thence into Persia, if, as appears to be the case, a Ghilan plant of Aucher Eloi (n. 5370) is the same species. Ledebour in his Flora Rossica, says that it inhabits the provinces of Mingrelia, Iberia, the Suwant, Lenkoran, and the south-west shores of the Caspian sea. It has been for some time known to amateurs in England, though not hitherto figured in any English work. One of its nearest allies is the very broad-leaved *C. byzantinum* (Tab. nost. 1122), which has a broader leaf, a much smaller paler flower, and broad short anthers; and is a native of Constantinople.

C. speciosum has been cultivated for many years in Kew, but the specimen here figured, which is much more deeply coloured than the Kew ones, was sent by Messrs. Barr and Sugden, who have a fine collection of the species of this beautiful genus.

DESCR. *Corm* the size of a walnut, clothed with rich chestnut-brown shining sheaves, of which one, as broad as the finger in diameter, extends four to five inches up the

JANUARY 1ST, 1874.

scapes. *Leafing-stem* one foot high, with three to five leaves. *Leaves* appearing before the flowers, a foot long, by two to four inches broad, elliptic, suberect, narrowed to the obtuse apex, of a dark green colour, paler beneath. *Flowers* numerous. *Perianth-tube* six to twelve inches long, as thick as a goose-quill, pale purple; limb four to five inches in diameter, of a clear red-purple colour with a white throat; segments elliptic, rounded at the point, concave, without conspicuous venation or tesselation. *Anthers* linear, oblong, yellow, bursting outwards. *Styles* three, subunilateral, slender, quite entire, white, tips slightly incurved. *Capsule* two-thirds inch long, turgid.—*J. D. H.*

TAB. 6079.

BAMBUSA STRIATA.

Native of China.

Nat. Ord. GRAMINIEÆ.—Tribe BAMBUSEÆ.

Genus BAMBUSA, *Linn.; (Munro in Trans. Linn. Soc.*, vol. xxvi. p. 87).

BAMBUSA *striata;* culmo gracili inermi, internodiis $\frac{1}{2}$-$\frac{3}{4}$ poll. diametro viridi aureoque striatis cavitate angusto, foliis 6–8-pollicaribus $\frac{3}{4}$–1 poll. latis, e basi obliqua obtusa elongato oblongo-lanceolatis acuminatis subtus subglaucescentibus glabris, ligula brevi truncata ciliata, vagina lævissima, paniculæ ramis gracillimis, spiculis angustis $\frac{3}{4}$ poll. longis subternatim fasciculatis, fasciculis remotis sessilibus, glumis oblongo-lanceolatis acuminatis compressis lævibus obscure 9–11-nerviis, inferiore longiore, palea inferiore subulato-lanceolata sub-enervi glaberrima, superiore paulo breviore angusta 2-nervi, nervis dorso ciliatis, squamulis 3 oblongis ciliatis, antheris 6 paleis æquilongis linearibus acuminatis rubris demum liliacinis, ovario styloque elongato piloso, stigmatibus 2 subulatis.

BAMBUSA striata, *Loddiges ex Lindl. in Penny Cyclopædia*, vol. iii. p. 357; *Munro Monog. Bambus. in Trans. Linn. Soc.*, vol. xxvi. p. 121.

The plant here figured has been, I believe, long known in this country as a native of China, and was introduced by the Messrs. Loddiges, of Hackney, many years ago. It is further cultivated in various tropical countries, and has been received at Kew both from the Jamaica and the Calcutta Botanic Gardens. In adopting the name *striata,* therefore, the only cause for hesitation is, that Lindley describes Loddiges' plant as having the leaves narrowed at the base, which hardly applies to this, in which they are oblique and almost rounded on the lower half of the base and acute in the upper. The specimen at Kew, sent from the Calcutta Botanic Gardens, is about six feet high, but Lindley describes it as attaining twenty feet, which from its habit it may very well be supposed to do. It belongs to Munro's third section of the genus *Bambusa,* which has a long hairy style, and to which the *B. vulgaris* and two other species belong.

JANUARY 1ST, 1874.

This plant flowered in November last, with Mr. Bull, who kindly sent me the specimen here figured; its anthers stain paper of a lilac colour; it has been called *B. Fortunei*, which I take to be a very different plant.

DESCR. A graceful tufted very glabrous slender species, six to twenty feet high. *Culms* as thick as the thumb; internodes four to six inches long, shining, striped yellow and green; walls thick, tube slender. Leaves six to eight inches by three-quarters to one inch long and broad, linear-oblong or oblong-lanceolate from an obtuse unequal base, glabrous, finely ciliolate on the margin, rather glaucous beneath; sheath slender, smooth, glabrous; ligula short, truncate, ciliate. *Panicle* slender, sparingly branched; branches long, with distant fascicles of about three sessile spikelets, which are three-quarters inch long, narrowly elliptic-lanceolate, 3–5-flowered. *Glumes* and lower *paleæ* similar, acuminate, with many obscure nerves, smooth; upper palea slender, 2-nerved, nerves ciliate. *Scales* 3, oblong, pilose. *Stamens* 6; anthers almost as long as the glumes, linear, apiculate, red purple, lilac when old. *Ovary* hairy, as is the very slender style; stigmas two.—*J. D. H.*

Fig. 1, Spikelet; 2, flower; 3, scales, base of filaments, and pistil:—*all magnified.*

COLONIAL AND FOREIGN FLORAS.

Flora Vitiensis; a Description of the Plants of the Viti or Fiji Islands, with an Account of their History, Uses, and Properties. By Dr. BERTHOLD SEEMANN, F.L.S. Royal 4to, 100 Coloured Plates, complete in one vol., cloth, £8 5s.

Flora of India. By Dr. J. D. HOOKER, F.R.S., and others. Part I., 10s. 6d.

Flora Capensis; a Systematic Description of the Plants of the Cape Colony, Caffraria, and Port Natal. By WILLIAM H. HARVEY, M.D., F.R.S., Professor of Botany in the University of Dublin, and OTTO WILHELM SONDER, Ph. D. Vols. I. and II. each 12s., Vol. III., 18s.

Flora of Tropical Africa. By DANIEL OLIVER, F.R.S., F.L.S. Vols. I. and II., each 20s. Published under the authority of the First Commissioner of Her Majesty's Works.

Flora Australiensis; a Description of the Plants of the Australian Territory. By G. BENTHAM, F.R.S., P.L.S., assisted by F. MUELLER, F.R.S., Government Botanist, Melbourne, Victoria. Vols. I. to VI., 20s. each. Published under the auspices of the several Governments of Australia.

Handbook of the New Zealand Flora; a Systematic Description of the Native Plants of New Zealand, and the Chatham, Kermadec's, Lord Auckland's, Campbell's, and Macquarrie's Islands. By Dr. J. D. HOOKER, F.R.S. Complete in one vol., 30s. Published under the auspices of the Government of that colony.

Flora of the British West Indian Islands. By Dr. GRISEBACH, F.L.S. 37s. 6d. Published under the auspices of the Secretary of State for the Colonies.

Flora Hongkongensis; a Description of the Flowering Plants and Ferns of the Island of Hongkong. By GEORGE BENTHAM, P.L.S. With a Map of the Island and Supplement by Dr. Hance, 18s. Published under the authority of Her Majesty's Secretary of State for the Colonies. The Supplement separately 2s. 6d.

Flora of Tasmania. By Dr. J. D. HOOKER, F.R.S. Royal 4to, 2 vols., 200 Plates, 17l. 10s. coloured. Published under the authority of the Lords Commissioners of the Admiralty.

On the Flora of Australia: its Origin, Affinities, and Distribution. By Dr. J. D. HOOKER, F.R.S. 10s.

Contributions to the Flora of Mentone, and to a Winter Flora of the Riviera, including the Coast from Marseilles to Genoa. By J. TRAHERNE MOGGRIDGE. Royal 8vo. Parts I. to IV. Each, with 25 Coloured Plates, 15s., or complete in one vol. 63s.

The Tourists' Flora; a Descriptive Catalogue of the Flowering Plants and Ferns of the British Islands, France, Germany, Switzerland, Italy, and the Italian Islands. By JOSEPH WOODS, F.L.S. 18s.

Outlines of Elementary Botany, as Introductory to Local Floras. By G. BENTHAM, F.R.S., President of the Linnean Society. Second Edition, 2s. 6d.

Laws of Botanical Nomenclature adopted by the International Botanical Congress, with an Historical Introduction and a Commentary. By ALPHONSE DE CANDOLLE. 2s. 6d.

L. REEVE AND CO., 5, HENRIETTA STREET, COVENT GARDEN.

NOW READY, Vol. VI., 20s.
FLORA AUSTRALIENSIS.
A Description of the Plants of the Australian Territory. By GEORGE BENTHAM, F.R.S., assisted by BARON FERDINAND MUELLER, C.M.G., F.R.S. Vol. VI. Thymeleæ to Dioscorideæ.

L. REEVE & Co., 5, Henrietta Street, Covent Garden.

NOW READY.
LAHORE TO YARKAND.
Incidents of the Route and Natural History of the Countries traversed by the Expedition of 1870, under T. D. FORSYTH, Esq., C.B. By GEORGE HENDERSON, M.D., F.L.S., F.R.G.S., Medical Officer of the Expedition, and ALLAN O. HUME, Esq., C.B., F.Z.S., Secretary to the Government of India. With 32 Coloured Plates of Birds and 6 of Plants, 26 Photographic Views of the Country, a Map of the Route, and Woodcuts. Price 42s.

L. REEVE & Co., 5, Henrietta Street, Covent Garden.

NOW READY.
Part X., completing the Work, with 10 Coloured Plates, Portrait and Memoir of the Author, and a full Index, price 25s.,
FLORA VITIENSIS,
A DESCRIPTION OF
THE PLANTS OF THE VITI OR FIJI ISLANDS,
WITH AN ACCOUNT OF
THEIR HISTORY, USES, AND PROPERTIES.
BY BERTHOLD SEEMANN, PH.D., F.L.S.
The Work complete in One Vol., with 100 Coloured Plates, £8 5s., cloth.

L. REEVE & CO., 5, HENRIETTA STREET, COVENT GARDEN.

NOW READY.
Demy 8vo, with 4 double Plates printed in Colours, 8 plain Plates, and five Woodcuts, price 10s. 6d.,
HARVESTING ANTS
AND
TRAP-DOOR SPIDERS.
NOTES AND OBSERVATIONS ON THEIR HABITS AND DWELLINGS.
BY J. TRAHERNE MOGGRIDGE, F.L.S.

L. REEVE & CO., 5, HENRIETTA STREET, COVENT GARDEN.

NOW READY, with 308 beautiful Wood Engravings, price 25s. (to Subscribers for the entire Work, 21s.), Vol. II. of
THE
NATURAL HISTORY OF PLANTS.
By Professor BAILLON,
President of the Linnean Society of Paris,
Translated, with Additional Notes and References,
By MARCUS M. HARTOG, B.Sc.

L. REEVE AND CO., 5, HENRIETTA STREET, COVENT GARDEN.

Third Series.

No. 350.

VOL. XXX. FEBRUARY. [*Price* 3s. 6d. col^d 2s. 6d. plain.

OR No. 1044 OF THE ENTIRE WORK.

CURTIS'S
BOTANICAL MAGAZINE,

COMPRISING

THE PLANTS OF THE ROYAL GARDENS OF KEW,

AND OF OTHER BOTANICAL ESTABLISHMENTS IN GREAT BRITAIN,
WITH SUITABLE DESCRIPTIONS;

BY

JOSEPH DALTON HOOKER, M.D., C.B., F.R.S., L.S., &c.

Director of the Royal Botanic Gardens of Kew.

Nature and Art to adorn the page combine,
And flowers exotic grace our northern clime.

LONDON:
L. REEVE & CO., 5, HENRIETTA STREET, COVENT GARDEN.
1874.

[*All Rights reserved.*]

NOW READY.
Demy 8vo, with 4 double Plates printed in Colours, 8 plain Plates, and five Woodcuts, price 10s. 6d.,

HARVESTING ANTS
AND
TRAP-DOOR SPIDERS.
NOTES AND OBSERVATIONS ON THEIR HABITS AND DWELLINGS.

By J. TRAHERNE MOGGRIDGE, F.L.S.

L. REEVE & CO., 5, HENRIETTA STREET, COVENT GARDEN.

BOTANICAL PLATES;
OR,
PLANT PORTRAITS.
IN GREAT VARIETY, BEAUTIFULLY COLOURED, 6d. and 1s. EACH.

List of nearly 2000, one stamp.

L. REEVE & Co., 5, Henrietta Street, Covent Garden.

FLORAL PLATES,
BEAUTIFULLY COLOURED BY HAND, 6d. EACH.

A New List of 500 Varieties, one stamp.

L. REEVE & Co., 5, Henrietta Street, Covent Garden.

NOW READY.
BENTHAM AND HOOKER'S
GENERA PLANTARUM.

Part IV. being the first part of Vol. II., comprising Caprifoliaceæ to Compositæ. Price 24s.

L. REEVE & Co., 5, Henrietta Street, Covent Garden.

"THE GARDEN,"

A Weekly Illustrated Journal devoted solely to Horticulture in all its branches.

THE GARDEN is conducted by WILLIAM ROBINSON, F.L.S., Author of "Hardy Flowers," "Alpine Flowers for English Gardens," "The Parks, Promenades, and Gardens of Paris," &c., and the best Writers in every department of Gardening are contributors to its pages.

The following are some of the subjects regularly treated of in its pages:—

The Flower Garden.	Hardy Flowers.
Landscape Gardening.	Town Gardens.
The Fruit Garden.	The Conservatory.
Garden Structures.	Public Gardens.
Room and Window Gardens.	The Greenhouse and
Notes and Questions.	The Household.
Market Gardening.	The Wild Garden.
Trees and Shrubs.	The Kitchen Garden.

THE GARDEN may be obtained through all Newsagents and at the Railway Bookstalls, at 4d. per copy. It may also be had direct from the Office at 5s. for a Quarter, 9s. 9d. for a Half-year, and 19s. 6d. for a Year, payable in advance; and in Monthly Parts. Specimen Copies (post free) 4½d.

37, Southampton Street, Covent Garden, W.C.

TAB. 6080.

FAGRÆA ZEYLANICA.

Native of Ceylon.

Nat. Ord. LOGANIACEÆ.—Tribe FAGRÆEÆ.

Genus FAGRÆA, *Thunb.; (DC. Prodr.*, vol. ix. p. 28).

FAGRÆA *zeylanica;* robusta, fruticosa, glaberrima, foliis 5-12-pollicaribus obovatis obovato-spathulatisve apice rotundatis sessilibus v. in petiolum robustum angustatis crasse coriaceis, nervis paucis inconspicuis, cymis terminalibus 3-multifloris, floribus breviter et crasse pedicellatis, calycis ovoidei lobis rotundatis scarioso-marginatis, bracteolis paucis brevibus, corollæ tubo 4-pollicari sursum infundibulari, lobis oblongis apice rotundatis, genitalibus exsertis, antheris magnis.

FAGRÆA zeylanica, *Thunb. Nov. Gen.*, vol. ii. p. 34; *Act. Holm.*, 1782, p. 125, t. 4; *Lamk. Ill.*, t. 167, f. 2; *Blume Rumphia*, vol. ii. t. 78; *Thwaites Enum.*, p. 200 et 425; *DC. Prodr.*, vol. ix. p. 29; *Benth. in Journ. Linn. Soc. Bot.*, 1857, p. 98.

SOLANDRA oppositifolia, *Moon Cat.*, p. 15.

A native of the central province of Ceylon, where, according to Dr. Thwaites, it abounds on the banks of the river at Balangodde. It is one of the handsomest species of a fine tropical Asiatic and Polynesian genus, of which some twenty species are enumerated by Bentham in his notes on *Loganiaceæ*, published in the *Linnæan Journal* (cited above), in 1857, to which several more are now to be added from the Malayan Islands. One species, the *F. obovata*, is figured in this work (Tab. 4205). The individual here figured was sent from Ceylon by Dr. Thwaites, about ten years ago, and flowered in the Royal Gardens in July, 1873.

DESCR. A very stout glabrous spreading deep green thick and coriaceous-leaved shrub. *Branches* as thick as the little finger, bright green, as are all the parts, except the flowers, scarcely shining. *Leaves* variable in length and breadth, five to twelve inches long, usually obovate, and narrowed into a very short, stout, semiterete petiole, which is, however, sometimes one to one and a half inches long; blade

FEBRUARY 1ST, 1874.

sometimes much elongated, and between obovate and spathulate; tip always rounded and sometimes emarginate; margins slightly recurved; nerves few, spreading, almost invisible in the fresh state. *Cymes* in terminal clusters, or solitary, three or more flowered, corymbose; peduncles and short pedicels as thick as a goosequill, smooth; bracts and bracteoles very short, triangular. *Calyx* nearly one inch long, ovoid, cleft to above the middle into rounded, appressed, overlapping lobes, with scarious margins when dry. *Corolla* white, leathery; tube four inches long, funnel-shaped above; limb three to four inches in diameter; lobes oblong, thick, spreading and recurved, nerveless. *Stamens* with very slender unequal filaments longer than the tube. *Anthers* nearly one-half inch long, broadly oblong, obtuse, pale yellow. *Ovary* cylindric; style slender, exserted; stigma green, capitate.—*J. D. H.*

Fig. 1, Specimen of *natural size*; 2, ovary; 3, transverse section of ovary; 4, top of filament and anther:—*all magnified*.

Tab. 6081.

GAILLARDIA Amblyodon.

Native of Texas.

Nat. Ord. Compositæ.—Tribe Helenioideæ.

Genus Gaillardia, *Foug.*; (*Benth. & Hook. f. Gen. Plant.*, vol. ii. p. 414).

Gaillardia *Amblyodon*; annua, ramosa, pilis patulis hirtella, foliis semi-amplexicaulibus oblongis v. inferioribus spathulatis apices versus grosse dentatis basi auriculato-bilobis, capitulis pedunculatis 2½ poll. latis, involucri squamis 3-4-seriatis lineari-lanceolatis setaceo-acuminatis hispidis conformibus externis basi concretis, receptaculi setis rigidis setaceo-acuminatis achænia superantibus, ligulis 12-15 obcuneato-oblongis trifidis sanguineo-purpureis, corollæ lobis extus pubescentibus, achæniis cylindraceo-oblongis, pappo radii e squamis brevibus latis exaristatis disci oblongo-oblanceolatis longe setaceo-acuminatis.

Gaillardia Amblyodon, *J. Gay in Ann. Sc. Nat.*, ser. 2, vol. ii. p. 57; *Torr. & Gray Fl. N. Am.*, vol. ii. p. 267; *Gray Chlor. Bot. Am.*, 32, t. 4.

A very handsome October-flowering annual, a native of sandy plains in Texas and New Mexico, where it blossoms from the beginning of summer until the winter's frost cuts it off. The genus to which it belongs inhabits both temperate North America and extra-tropical South America, and consists of about eight species, of which two have been already figured in this Magazine—namely, the old *G. bicolor*, Lam. (Tab. 1602), and its variety *Drummondii* (Tabs. 3368 and 3551); and the large yellow-flowered *G. aristata* (Tab. 2940). The present species was discovered by Berlandier in 1827, and collected subsequently by Lindheimer in 1844, and by Drummond in 1845. The specimen here figured was raised from seed by Mr. Thompson of Ipswich, and flowered in October, 1873.

Descr. An annual branching herb, two to three feet high, forming large clumps, clothed with spreading short hispid hairs. *Leaves*, radical subspathulate; cauline one and a half to two and a half inches long, semi-amplexicaul, oblong, subacute, coarsely toothed beyond the middle, usually contracted below it; pubescent and pale beneath, midrib beneath hispid, base 2-lobed, auricled. *Heads* two and a half inches diameter, terminal, peduncled. *Involucral scales* green, rigid, 3–4-seriate, subulate-lanceolate, hispid, erecto-patent, the outer confluent at the base. *Receptacle* clothed with rigid bristles which exceed the achenes. *Ray-flowers* twelve to fourteen, spreading, deep blood-red; limb cuneate-oblong, obtusely 3-lobed. *Disk-flowers* short; lobes short, obtuse, erect, pubescent externally. *Achenes* cylindric-oblong, those of the ray with a pappus of few short broad scales; those of the disk with as many long rigid scales that terminate in setaceous points; style-arms long, slender, exserted.—*J. D. H.*

Fig. 1, Ray-flower; 2, disk-flower; 3, pappus scale of the latter:—*all magnified.*

TAB. 6082.

STAPELIA CORDEROYI.

Native of South Africa.

Nat. Ord. ASCLEPIADEÆ.—Tribe STAPELIEÆ.

Genus STAPELIA, *Linn.*; (*Decaisne in DC. Prodr.*, vol. viii. p. 652).

STAPELIA (Duvalia) *Corderoyi*; humilis, glaberrima, glauca, ramulis brevibus obesis procumbentibus ovoideo-oblongis obtusis sub-4-costatis, costis rotundatis remote dentatis, sinubus acutis, dentibus brevibus triangulari-subulatis patentibus basi carnosis et utrinque unituberculatis, corolla 1½-2 poll. lata ad medium 5-loba, lobis triangularibus acuminatis sordide viridibus marginibus recurvis, apices versus fusco-purpurascentibus, sinubus setoso-glandulosis, fauce elevata pallide-lilacina filamentosa, corona staminea duplici breviter stipitata purpurea, exteriore e disco crasso obtuse 5-gono, interiore e cornubus 5 brevibus crassis ovoideis exteriore impositis.

I am quite unable to identify this very curious little *Stapelia* with any described species, though it clearly belongs to Haworth's section *Duvalia*. In habit, size, and form of branches it agrees with *S. cæspitosa*, Mass., but the flowers are very much larger, and of a totally different form and colour. To the same division belong *S. radiata* (Tab. nost. 619) and *S. reclinata* (Tab. nost. 1397); but these have, like *S. cæspitosa*, small dark-coloured flowers, with very narrow corolla-lobes.

I have named this very curious and distinct species after Mr. Justus Corderoy, of Blewberry, near Didcot, an old and an eminent cultivator of succulent plants, and for many years a valued correspondent of the Royal Gardens. It flowered at Blewberry in September of last year.

DESCR. *Branches* short, procumbent, very stout, glaucous, about two inches long by three-quarters of an inch in diameter, very pale green and fleshy, obtusely 4-5-ribbed, the ribs semi-cylindric with an acute sinus between them, each bearing two to four short triangular teeth, which are fleshy at the base, and there furnished with a globose tubercle on each side. *Peduncles* solitary or in pairs, about an inch long,

FEBRUARY 1ST, 1874.

green, variegated with purple. *Calyx* of five triangular-subulate green teeth, with red-brown tips. *Corolla* about one and a half inches in diameter, 5-lobed to about the middle; lobes triangular-acuminate, dirty green, with purple brown tips, and a few long slender glandular purple hairs in the sinus, margin recurved; throat surrounded with an elevated lilac coronal disk, clothed with slender spreading purplish hairs. *Staminal-column* on a short stipes, expanded into a broad fleshy purple obscurely 5-lobed disk (the outer corona) which bears on its summit as many egg-shaped obtuse spreading horns (which form the inner corona). *Pollen-masses* bright orange, reniform.—*J. D. H.*

Fig. 1, Teeth on ribs of branches; 2, flower, with the corolla removed; 3, pollen-masses:—*all magnified.*

Tab. 6083.

IRIS Douglasiana.

Native of California.

Nat. Ord. Iridaceæ.—Tribe Irideæ.

Genus Iris, *Linn.*; (*Endl. Gen. Plant.*, p. 166).

Iris *Douglasiana;* imberbis, rhizomate crassitie digitis, foliis 1–1½-pedalibus ½–¾ poll. latis planis anguste lineari-ensiformibus longe acuminatis scapum solidum excedentibus, spathæ valvis 2–3-pollicaribus angustis acuminatis pedunculos longe superantibus, ovario angusto obtuse 3-gono faciebus concavis, perianthii tubo ½–¾ pollicari, limbi 3–4 poll. lati segmentis exterioribus obovato-spathulatis pallide lilacinis disco albido venis purpureis, interioribus erectis elliptico-lanceolatis acuminatis lilacinis, stigmatibus cuneato-oblongis 2-fidis, lobis acutis dentatis.

Iris Douglasiana, *Herbert in Hook. & Arn. Bot. Beech. Voy.* 395; *Torrey in Whipple Rep. Bot.* 35th *Parallel*, p. 144.

Discovered by Coulter in California, and subsequently collected by David Douglas, in 1833, in New California, but unknown to me from any other locality and collector, except from a mention of the plant in one of the Reports of the United States' surveys, quoted above, where it is stated to be found on hill-sides in the Grass Valley, California, together with a large-flowered variety (how large it is not said), and longer pedicels (one inch) at the Corte Madera, also in California. It is a very little known plant, being omitted in Klatt's monograph of the genus, (published in the Linnæ, vol. xxxiv.), and is closely allied to *S. longipetala* (Tab. nost. 5298), which is, however, a very much larger species, with a remarkably short perianth-tube. I am indebted to Messrs. Veitch for the specimen here figured, which flowered in his nursery last year.

Descr. *Rhizome* as thick as the little finger, creeping. *Leaves* a foot to a foot and a half long, by half to three-quarters of an inch in diameter, of a dark green colour, except at the bases and on the sheaths, which are paler, variegated with red, narrow-linear, gradually contracted into the acuminate tip, nerves obscure. *Spathes* usually two, enclosing together two flowers, three to four inches long, narrow and long acuminate, without scarious margins. *Peduncles* shorter than the ovary, which is one to one and a half inches long, narrow-oblong, with three rounded angles and channelled faces. *Perianth-tubes* one-half to three-quarters of an inch long, rather stout, green; limb three to four inches in diameter; outer segments obovate-spathulate, spreading and recurved, beardless, obtusely toothed, pale lilac, with a white disk which is veined with purple; inner segments rather shorter, lanceolate, acuminate, erect; obtusely toothed, pale lilac-purple, not veined. *Stigmas* one half as long as the inner segments, bifid, oblong cuneate, segments acute.—*J. D. H.*

Tab. 6084.

ODONTOGLOSSUM roseum.

Native of Peru.

Nat. Ord. Orchideæ.—Tribe Vandeæ.

Genus Odontoglossum, *H. B. & K.; (Lindl. fol. Orchid. Odontoglossum).*

Odontoglossum *roseum*; pseudobulbis late-ovatis compressis ancipitibus apice 2-foliatis, foliis loriformibus acutis canaliculatis sessilibus dorso carinatis coriaceis enerviis, racemis breviuscule pedunculatis cernuis elongatis multifloris, rachi tenui, bracteis ovato-lanceolatis viridibus pedicellis æquilongis, ovario gracili perianthio roseo 1-poll. diametro, sepalis petalisque consimilibus oblongo-ellipticis subacutis patenti-recurvis, labello anguste 3-lobo, lobis lateralibus brevibus rotundatis, intermedio longe producto lineari apice paulo dilatato obtuso, disco inter lobos laterales callo 4-fido ornato, columna apice pallida membranacea 3-fida.

Odontoglossum roseum, *Lindl. in Benth. Plant. Hartweg,* p. 151; *Fol. Orchid. Odontogloss.* p. 23: *Reichb. f. in Walp. Ann.,* vol. vi. t. 848; *Gard. Chron.,* 1867, p. 404; *André in Linden Illust. Hortic.,* vol. xviii. t. 66; *Bateman Monog. of Odontogloss.* t. 22.

In its rose-coloured flowers this forms a remarkable contrast to the prevalent colour of the genus to which it belongs. It was discovered by Hartweg near Loxa, in the Peruvian Andes, in a quite cool region, and was introduced by Mr. Linden from that region by his able collector, Mr. Wallis, in 1865. A figure of a small and poor specimen is given in Mr. Bateman's beautiful work upon this genus, and a much finer one in the "Illustration Horticole," where, however, the flowers are represented as larger and of a much deeper hue than in our plant. The specimen here figured was exhibited by Mr. Linden at the Horticultural Society in April, 1871.

Descr. *Pseudobulbs* two inches long by one and a half in diameter, broadly ovate, much compressed, 2-edged, pale-brown when mature. *Leaves* five to seven inches long, by three-quarters to one inch in diameter, strap-shaped, acute, narrowed at the base, channelled and dark-green above, paler and keeled beneath, very coriaceous, nerveless. *Racemes* proceeding from the base of the pseudobulb, six to eight inches long, on a peduncle of half that length, graceful, curved, many-flowered; peduncle closely clothed with appressed obtuse sheathing green bracts; rachis rather flexuous; flowers rather distant and distichous; bracts about as long as the pedicels, acute, appressed, green. *Ovary* very slender, 3-angled, hardly distinguishable from the pedicel. *Perianth* one to one and a quarter inches in diameter, rose-red throughout. *Sepals* and *petals* similar, oblong-elliptic, acute, spreading and recurved, paler at the back. *Lip* about as long as the petals; claw appressed to the column; limb cuneate at the base, 3-lobed; lateral lobes small, rounded, enclosing a small disk which bears a 4-lobed appressed callus; midlobe much longer than the rest of the lip, linear, dilated or spathulate at the obtuse tip, obscurely channelled above. *Column* rather slender, rose-coloured, with a 3-toothed white membranous tip.—*J. D. H.*

Fig. 1, View of column and lip from above; 2, the same from the disk:—*both magnified*.

COLONIAL AND FOREIGN FLORAS.

Flora Vitiensis; a Description of the Plants of the Viti or Fiji Islands, with an Account of their History, Uses, and Properties. By Dr. BERTHOLD SEEMANN, F.L.S. Royal 4to, 100 Coloured Plates, complete in one vol., cloth, £8 5s.

Flora of India. By Dr. J. D. HOOKER, F.R.S., and others. Part I., 10s. 6d.

Flora Capensis; a Systematic Description of the Plants of the Cape Colony, Caffraria, and Port Natal. By WILLIAM H. HARVEY, M.D., F.R.S., Professor of Botany in the University of Dublin, and OTTO WILHELM SONDER, Ph. D. Vols. I. and II. each 12s., Vol. III., 18s.

Flora of Tropical Africa. By DANIEL OLIVER, F.R.S., F.L.S. Vols. I. and II., each 20s. Published under the authority of the First Commissioner of Her Majesty's Works.

Flora Australiensis; a Description of the Plants of the Australian Territory. By G. BENTHAM, F.R.S., P.L.S., assisted by F. MUELLER, F.R.S., Government Botanist, Melbourne, Victoria. Vols. I. to VI., 20s. each. Published under the auspices of the several Governments of Australia.

Handbook of the New Zealand Flora; a Systematic Description of the Native Plants of New Zealand, and the Chatham, Kermadec's, Lord Auckland's, Campbell's, and Macquarrie's Islands. By Dr. J. D. HOOKER, F.R.S. Complete in one vol., 30s. Published under the auspices of the Government of that colony.

Flora of the British West Indian Islands. By Dr. GRISEBACH, F.L.S. 37s. 6d. Published under the auspices of the Secretary of State for the Colonies.

Flora Hongkongensis; a Description of the Flowering Plants and Ferns of the Island of Hongkong. By GEORGE BENTHAM, P.L.S. With a Map of the Island and Supplement by Dr. Hance. 18s. Published under the authority of Her Majesty's Secretary of State for the Colonies. The Supplement separately 2s. 6d.

Flora of Tasmania. By Dr. J. D. HOOKER, F.R.S. Royal 4to, 2 vols., 200 Plates, 17l. 10s. coloured. Published under the authority of the Lords Commissioners of the Admiralty.

On the Flora of Australia: its Origin, Affinities, and Distribution. By Dr. J. D. HOOKER, F.R.S. 10s.

Contributions to the Flora of Mentone, and to a Winter Flora of the Riviera, including the Coast from Marseilles to Genoa. By J. TRAHERNE MOGGRIDGE. Royal 8vo. Parts I. to IV. Each, with 25 Coloured Plates, 15s., or complete in one vol. 63s.

The Tourists' Flora; a Descriptive Catalogue of the Flowering Plants and Ferns of the British Islands, France, Germany, Switzerland, Italy, and the Italian Islands. By JOSEPH WOODS, F.L.S. 18s.

Outlines of Elementary Botany, as Introductory to Local Floras. By G. BENTHAM, F.R.S., President of the Linnean Society. Second Edition, 2s. 6d.

Laws of Botanical Nomenclature adopted by the International Botanical Congress, with an Historical Introduction and a Commentary. By ALPHONSE DE CANDOLLE. 2s. 6d.

L. REEVE AND CO., 5, HENRIETTA STREET, COVENT GARDEN.

NOW READY, Part II., 10s. 6d.

FLORA OF INDIA.

BY

DR. HOOKER, C.B., F.R.S.

Assisted by various Botanists.

L. REEVE & Co., 5, Henrietta Street, Covent Garden.

NOW READY, Vol. VI., 20s.

FLORA AUSTRALIENSIS.

A Description of the Plants of the Australian Territory. By GEORGE BENTHAM, F.R.S., assisted by BARON FERDINAND MUELLER, C.M.G., F.R.S. Vol. VI. Thymeleæ to Dioscorideæ.

L. REEVE & Co., 5, Henrietta Street, Covent Garden.

NOW READY.

LAHORE TO YARKAND.

Incidents of the Route and Natural History of the Countries traversed by the Expedition of 1870, under T. D. FORSYTH, Esq., C.B. By GEORGE HENDERSON, M.D., F.L.S., F.R.G.S., Medical Officer of the Expedition, and ALLAN O. HUME, Esq., C.B., F.Z.S., Secretary to the Government of India. With 32 Coloured Plates of Birds and 6 of Plants, 26 Photographic Views of the Country, a Map of the Route, and Woodcuts. Price 42s.

L. REEVE & Co., 5, Henrietta Street, Covent Garden.

NOW READY.

Part X., completing the Work, with 10 Coloured Plates, Portrait and Memoir of the Author, and a full Index, price 25s.,

FLORA VITIENSIS,

A DESCRIPTION OF

THE PLANTS OF THE VITI OR FIJI ISLANDS,

WITH AN ACCOUNT OF

THEIR HISTORY, USES, AND PROPERTIES.

By BERTHOLD SEEMANN, PH.D., F.L.S.

The Work complete in One Vol., with 100 Coloured Plates, £8 5s., cloth.

L. REEVE & CO., 5, HENRIETTA STREET, COVENT GARDEN.

NOW READY, with 308 beautiful Wood Engravings, price 25s. (to Subscribers for the entire Work, 21s.), Vol. II. of

THE

NATURAL HISTORY OF PLANTS.

By Professor BAILLON,

President of the Linnean Society of Paris,

Translated, with Additional Notes and References,

By MARCUS M. HARTOG, B.Sc.

L. REEVE AND Co., 5, HENRIETTA STREET, COVENT GARDEN.

Third Series.

No. 351.

VOL. XXX. MARCH. [*Price 3s. 6d. col*^d *2s. 6d. plain.*]

OR No. 1045 OF THE ENTIRE WORK.

CURTIS'S
BOTANICAL MAGAZINE,

COMPRISING

THE PLANTS OF THE ROYAL GARDENS OF KEW,

AND OF OTHER BOTANICAL ESTABLISHMENTS IN GREAT BRITAIN,
WITH SUITABLE DESCRIPTIONS;

BY

JOSEPH DALTON HOOKER, M.D., C.B., F.R.S., L.S., &c.

Director of the Royal Botanic Gardens of Kew.

Nature and Art to adorn the page combine,
And flowers exotic grace our northern clime.

LONDON:
L. REEVE & CO., 5, HENRIETTA STREET, COVENT GARDEN.
1874.

[*All rights reserved.*]

ROYAL BOTANIC SOCIETY OF LONDON,
GARDENS—REGENT'S PARK.

ARRANGEMENTS FOR 1874.

EXHIBITIONS OF SPRING FLOWERS, Wednesdays, March 25, April 22.
SUMMER EXHIBITIONS, Wednesdays, May 20, June 10, and June 24. Gates open at 2 o'clock.
SPECIAL EVENING FETE, Wednesday July 8. Gates open at 8 o'clock P.M. Evening Dress.
AMERICAN EXHIBITION, Daily, May 25, to June 9.
PROMENADES.—Every Wednesday in May, June, and July, excepting the Exhibition days, commencing May 6. Visitors admitted only by the Special Coloured Orders.
LECTURES in the Museum at 4 o'clock precisely, Fridays May 15, 22, 29; June 5, 12, 19, 26; July 3.

ROYAL HORTICULTURAL SOCIETY.

MEETINGS AND SHOWS IN 1874.

March	4.	(Fruit and Floral Meeting.)
,,	18.	(Hyacinths.)
April	1.	(Fruit and Floral Meeting.)
,,	15.	(Early Rhododendrons.)
May	13.	(Pot Roses.)
,,	27.	(Fruit and Floral Meeting.)
June 4 and 5.		(Great Summer Show.)
,,	17.	(Fruit and Floral Meeting.)
July	1.	(Cut Roses.)
,,	15.	(Zonal Pelargoniums.)
August	5.	(Fruit and Floral Meeting.)
,,	19.	Do.
September	2.	(Dahlias.)
October	7.	(Fruit and Floral Meeting.) (Fungi.)
November	11.	(Fruit and Chrysanthemums.)
December	2.	(Fruit and Floral Meeting.)

"THE GARDEN,"

A Weekly Illustrated Journal devoted solely to Horticulture in all its branches.

THE GARDEN is conducted by WILLIAM ROBINSON, F.L.S., Author of "Hardy Flowers," "Alpine Flowers for English Gardens," "The Parks, Promenades, and Gardens of Paris," &c., and the best Writers in every department of Gardening are contributors to its pages.

The following are some of the subjects regularly treated of in its pages:—

- The Flower Garden.
- Landscape Gardening.
- The Fruit Garden.
- Garden Structures.
- Room and Window Gardens.
- Notes and Questions.
- Market Gardening.
- Trees and Shrubs.
- Hardy Flowers.
- Town Gardens.
- The Conservatory.
- Public Gardens.
- The Greenhouse and The Household.
- The Wild Garden.
- The Kitchen Garden.

THE GARDEN may be obtained through all Newsagents and at the Railway Bookstalls, at 4d. per copy. It may also be had direct from the Office at 5s. for a Quarter, 9s. 9d. for a Half-year, and 19s. 6d. for a Year, payable in advance; and in Monthly Parts. Specimen Copies (post free) 4½d.

37, Southampton Street, Covent Garden, W.C.

TAB. 6085.

ODONTOGLOSSUM ROEZLII.

Native of New Grenada.

Nat. Ord. ORCHIDEÆ.—Tribe VANDEÆ.

Genus ODONTOGLOSSUM, *H. B. & K.*; (*Lindl. Fol. Orchid., Odontoglossum*).

ODONTOGLOSSUM *Roezlii*; pseudobulbis parvis anguste ovatis compressis marginibus acutis, foliis pedalibus elongatis lineari-lanceolatis acuminatis carinatiset lineato-nervoris, scapis gracilibus foliis brevioribus 1-2 floris, floribus maximis, perianthio plano, sepalis obovato-oblongis acutis niveis, petalis sepalis consimilibus niveis fascia magna lata versus basin sanguineo-purpurea ornatis, labello maximo late obcordato in sinu apiculato, ima basi in unguem brevem contracto, ungue utrinque postice in spinam rectam producto, disco 5-carinato spinisque aureo rubroque irroratis, columna breviuscula exalata.

ODONTOGLOSSUM Roezlii, *Reichb. f. in Gard. Chron.*, 1873, p. 1303, *cum Ic. Xylog., et in Xen. Orchid.*, vol. ii. p. 191 (*cum. tab.* 182 *ined.*)

This is a very near ally and a rival of the *O. vexillarium* (Tab. nost. 6037); so near an ally indeed, that Prof. Reichenbach suggests the possibility of its being a hybrid between that plant and *O. Phalænopsis*. Putting aside the different colour of the flower, the principal distinctions between this and *O. vexillarium* are the more slender leaves, which are nerved beneath, less robust habit, fewer-flowered scapes, the obcordate lip and longer column of this; Prof. Reichenbach indicates the flat (not revolute) sepals and the different keels at the base of the lip (he, however, finds three keels in this, not five, as in our specimen); to these may be added the more slender and much longer floral bracts, and shorter scapes. It is stated to be a native of New Grenada, where it was discovered by M. Roezl, whose name it bears; and was flowered by Mr. Bull in October last; to whom I am indebted for the opportunity of figuring it. It is a superb plant, and in respect of the pearly whiteness of the flower more admired by some than even *O. vexillarium*.

DESCR. *Pseudobulbs* one to two inches long, narrowly ovate, compressed. *Leaves* eight to twelve inches long, narrowly linear-lanceolate, acuminate, pale clear green above, paler and keeled beneath; nerves parallel. *Scapes* about half the length of the leaves, slender, terete, 1–2-flowered; bracts subulate-lanceolate, half an inch long, green; pedicels exceeding the bracts, gradually passing into the slender grooved ovary. *Flowers* three to three and a half inches across, but probably variable in size, almost as large as and closely resembling in form those of *O. vexillarium*; perianth quite flat. *Sepals* subequal, one and a half inches long, the dorsal rather the narrowest, obovate-oblong, acute, snow-white. *Petals* as large as, and altogether similar to, the lateral sepals, but with a broad red-purple band across their breadth towards the base. *Lip* very large, two to two and a quarter inches in breadth, broadly obcordate with a mucro in the notch, very shortly clawed, snow-white, with faintly yellow marblings, tinged with pale red on the disk above the base; a small spur-like horn arises on each side of the base of the claw, and is directed upwards and backwards, one on each side of the column; and there are five short slender ridges on the disk in front of the interval between the spurs. *Column* not winged.—*J. D. H.*

Fig. 1, Front; and 2, side view of portion of lip and column:—*magnified*.

TAB. 6086.

BAUHINIA NATALENSIS.

Native of Natal.

Nat. Ord. LEGUMINOSÆ.—Tribe BAUHINIEÆ.

Genus BAUHINIA, *Linn.*; (*Benth. & Hook. f. Gen. Pl.*, vol. i. p. 575).

BAUHINIA (Pauletia) *natalensis*; frutex inermis, erectus, glaberrimus, ramulis gracilibus, foliis parvis gracile petiolatis, foliolis 2 liberis oblique oblongis v. obovato-oblongis apice rotundatis, basi obtusis, pedunculis 1-2-floris oppositifoliis gracilibus, stipulis setiformibus, floribus 1½ poll. diam. erectis albis, calyce spathaceo late cymbiformi apiculato, petalis obovatis apice rotundatis, staminibus 5 longioribus filamentis 2 basin versus calcaratis, 5 duplo minoribus antheris parvis, ovarii stipite libero, stylo elongato, stigmate capitato, legumine plano acinaciformi acuminato glaberrimo tenuiter venoso, basin versus sensim angustato, margine inferiore plano.

BAUHINIA natalensis, *Oliv. Mss. in Herb. Kew.*

My first knowledge of this elegant little shrub was derived from specimens collected in Natal by Mr. Moodie, and communicated by Mr. McKen, the late energetic Curator of the D'Urban Botanic Gardens, in 1869. These were followed by pods with ripe seeds in 1870, from which the plant here figured was raised, and which flowered for the first time in September last. It is closely allied to the African and Indian *B. tomentosa*, Linn. (Tab. nost. 5560) and especially to a nearly glabrous and small-leaved variety of that plant from Port Natal, but the leaflets are perfectly free, the flowers much smaller and the stamens quite different.

DESCR. A small, glabrous, slender, leafy bush. *Branchlets* nearly straight, slender. *Leaves* alternate, somewhat distichous; petiole very slender, quarter to a half inch long, ending in a subulate point between the leaflets, swollen at the base; leaflets one inch long, quite free, obliquely obovate, or subovate-oblong, rounded at the apex, as also at the base on the outer side, dark green, rather paler beneath; midrib and few nerves very slender; stipules subulate. *Peduncles* leaf-opposed, 1-2-flowered, with two minute setaceous bracts

MARCH 1ST, 1874.

at the base. *Flower* one and a half inches in diameter, pure white with a faint crimson streak along the midrib of the three smaller petals. *Calyx* with a short turbinate tube, and a broad spathaceous green apiculate limb, one-third inch long, which is truncate at the base. *Petals* erecto-patent, obovate-oblong, rounded at the apex, obscurely veined; three upper rather smaller. *Stamens* ten; filaments almost free, and slightly hairy at the base; five longer equalling the style, of which two have each a lateral spur above the base; five shorter stamens half as long as the others; anthers all oblong, obtuse, yellow; cells ciliate at the base. *Ovary* stipitate, free, slender; style stout, elongate, stigma large, capitate. *Pod* three inches long by nearly one half inch broad, scimitar-shaped, acuminate, contracted at the base, flat, glabrous, obscurely reticulately nerved, convex edge flat, about 6-seeded, interior almost divided between the seeds by a thin down.—*J. D. H.*

Fig. 1, Leaf and portion of stem; 2, flower, with the petals removed; 3, long and short stamens:—*all magnified.*

TAB. 6087.

ARABIS BLEPHAROPHYLLA.

Native of California.

Nat. Ord. CRUCIFERÆ.—Tribe ARABIDEÆ.

Genus ARABIS, *Linn.*; (*Benth. & Hook. f. Gen. Pl.*, vol. i. p. 69).

ARABIS *blepharophylla*; perennis, erecta, caulibus foliosis, foliis ciliatis et sparse pilosis radicalibus rosulatis obovato-spathulatis obtusis sinuato-v. serrato-dentatis, caulinis elliptico-v. lineari-oblongis obtusis basi simplicibus v. subauriculatis, racemis brevibus latis obtusis, floribus gracile pedicellatis amplis roseis, petalis obovato-cordatis, siliquis $1\frac{1}{4}$-pollicaribus erectis rectis v. lente curvis linearibus, valvis utrinque obtusis costa nervisque lateralibus flexuosis validis, stylo brevissimo, seminibus 1-seriatis orbicularibus compressissimis exalatis brunneis.

ARABIS blepharophylla, *Hook. et Arn. Bot. Beech. Voy.*, p. 321.

Of the large genus *Arabis* almost all have white flowers; in a very few species they are yellow, and in this alone of those known to me, do the colour and size of flower together recommend it for cultivation. It is a native of San Francisco, in California, where it was discovered by David Douglas in 1833, and has since been collected by Bridges, Brewer, Bolander, and others, and is described as a great ornament in March on the hills of that State. It seems remarkable that so conspicuous a plant, growing in what is now a populous State, should be so little known, but I find no other description of it than that in the Botany of Beechey's Voyage, published thirty-five years ago; nor do I find any mention of it in the multitudinous and cumbrous records of the United States' Surveying Expeditions. Professor Asa Gray, of Cambridge, was, I believe, the first to send ripe seeds to England—this was in 1865—from which plants were raised at Kew, and by Mr. Thompson, of Ipswich, if I recollect aright; but it was not till quite recently that the plants throve (from seeds sent by Commissioner Watt, of the Agricultural Department of Washington) and appeared in their full beauty. The specimen here figured flowered at Kew in January, in a cool

MARCH 1ST, 1874.

frame, where it has hitherto thriven better than in the open border or rockwork; it is, however, doubtless quite hardy, and would succeed equally well out of doors, where, from its beauty and early flowering, it is sure to become a great favourite.

DESCR. Whole plant six to ten inches high, erect. *Root* perennial, fusiform. *Flower-stem* leafy, robust. *Leaves* all ciliate, and sparsely hairy with long simple or forked hairs; radical forming a lax rosette three to four inches in diameter, spreading, one to two and a half inches long, petioled, obovate-spathulate, obtuse, irregularly sinuate or toothed, dark green above, paler beneath; cauline leaves shorter, sessile, linear-oblong, obtuse, serrate or toothed, base rounded or slightly auricled. *Flowering-racemes* about two inches long, and nearly as broad, rounded at the apex; pedicels half an inch long, slender, spreading, erect in fruit. *Flower* three-quarters of an inch in diameter. *Sepals* erect, linear-oblong, obtuse. *Petals* with a short claw and broadly obovate retuse rose-coloured limb. *Filaments* slender; anthers small. *Pod* one to one and a half inches long, nearly one-eighth of an inch broad, erect, slightly curved, linear; valves obtuse at both ends, rather coriaceous, margined, convex over the seeds; midrib and waved lateral nerves very strong, giving the surface a grooved appearance; style very short, conical; stigma minute. *Seeds*, about eight to ten in each cell, 1-seriate, orbicular, much compressed, brown, not winged.— *J. D. H.*

Fig. 1, Flower with calyx and anthers removed; 2, ovary, 3, immature capsule:—*all magnified*.

TAB. 6088.

NUNNEZHARIA (Chamædorea, *Auct.*)
GEONOMÆFORMIS.

Native of Guatemala.

Nat. Ord. Palmeæ.—Tribe Arecineæ.

Genus Nunnezharia; *Ruiz & Pav.*—(Chamædorea, *Willd. et auct.*).

Nunnezharia (Psilostachys) *geonomæformis*; caudice gracili erecto dense annulato, foliis erecto-patentibus breviter petiolatis simplicibus obovatis apice bipartitis, vaginis brevibus apertis, spadicibus ♂ infra et inter coronam enatis longe pedunculatis, pedunculo erecto, masculi ramis gracillimis pendulis densifloris, fl. ♂ compresso-globosis, perianthio exteriore annulari brevissimo latissime 3-lobo, interiore e foliolis 3 obovato-triangularibus, antheris inclusis oblongis, fl. ♀ scrobiculis spadicis suberecti subimmersis depresso-globosis, perianthii foliolis 3 exterioribus interiora rotundata ovarium amplectentia subæquantibus, staminodiis minutissimis, stigmatibus minutissimis exsertis.

Chamædorea geonomæformis, *Wendland Ind. Palm.*, p. 12 (1854), *et in Otto et Dietr. Gartenz.*, 1852, *ex Oersted in Palm. Centroamer. in Natur. Foren. Vidensk. Meddels.*, 1858, p. 24; *Oersted, L'Ameriq. Centr.* Fasc. i. p. 14, t. 5 (1853).

Ch. fenestrata, *Hort. Houtt., ex Wendl. Ind. Palm*, 12.

Ch. humilis, *Hort. Berlin, ex Wendl. l.c.*

This little Palm was received at Kew from the Royal Gardens of Berlin in 1856, and flowered in the Palm-house in May, 1859, and repeatedly since. From its dwarf habit, abundant foliage, and graceful male inflorescence, it is one of the most elegant of the beautiful genus to which it belongs. It is a native of Guatemala, whence it was introduced by Warsiewicz, and named by Wendland. Its foliage precisely accords with that of a Peruvian congener, the *N.? geonomoides*, Spruce (Journ. Linn. Soc., vol. xi. p. 122); but the flowers are very much larger and widely different. There is no question but that, as Mr. Spruce points out, Willdenow's generic name of *Chamædorea* must give place to the prior one

March 1st, 1874.

of *Nunnezharia*, given to the genus nine years earlier by Ruiz and Pavon.

The Kew plant, which in 1858 (when the accompanying drawing was made), had a stem only a few inches high, with four naked joints, has now a stem three and a half feet high, which presents sixty-four joints between the rootlets and lowest leaf base. It is stated to have borne sometimes male and sometimes female spadices.

DESCR. Whole height about four feet. *Stem* erect, as thick as the thumb, deep bright green; internodes one-half to one inch long, not much contracted at the middle. *Leaves* spreading, eight to twelve inches long by five or six broad, obovate, obscurely serrate, apex two-partite, with spreading triangular lobes, deep green, plaited; nerves about twelve on each side, perfectly glabrous; petiole short, green; sheath oblong, the lower pale red-brown. *Spadices* (male) axillary, and from the joints immediately below the leaves, very slender, erect, terminated by long slender alternate drooping branches, eight to ten inches long; peduncle clothed with slender, erect, orange-brown, acuminate sheaths four inches long; branches very graceful, green, clothed throughout with close-set but not crowded male flowers. *Flowers* (male) compressed-globose, a quarter of an inch in diameter, dark green like the branch of the spadix, in which their bases are hardly sunk. *Outer perianth* of three minute membranous segments, connate into a cup; inner much larger, obovate, connate at the tips for some time. *Stamens* six, surrounding a rudimentary ovary. *Female flower* (from Oersted's description) immersed in pits of the erect branches of the spadix. *Outer perianth* nearly as large as the inner. *Staminodes* very minute.—*J. D. H.*

Fig. 1, Reduced view of whole plant; 2, male spadix:—*of the natural size;* 3, portion of ditto and flower; 4 male flower:—*magnified.*

TAB. 6089.

RHIPSALIS Houlletii.

Native of Brazil?

Nat. Ord. CACTEÆ.—Tribe OPUNTIEÆ.

Genus RHIPSALIS, *Gærtn.; (Benth. & Hook. f. Gen. Plant.*, vol. i. p. 850).

RHIPSALIS *Houlletii;* epiphytica, pendula, ramosa, glaberrima, caulibus gracilibus, ramulis foliaceo-dilatatis planis, internodiis elliptico-lanceolatis 1–1¼-poll. diam. grosse obtuse serratis coriaceo-carnosis nervis obscuris, floribus fere 1-poll. diametro pallide flavis odoris, ovario exserto oblongo obtuse 4-5-costato, perianthii foliolis 8-12 erecto-patentibus lanceolatis acutis, exterioribus paullo minoribus, staminibus numerosis perianthio brevioribus, stylo gracili, stigmatibus 4-5.

RHIPSALIS Houlletii, *Lemaire, Les Cacteæ*, p. 80, *nomen tantum*.

This *Rhipsalis* has been cultivated for some time in the Royal Gardens, where it flowered first in November, 1872, and it has been received also from Mr. Corderoy, who sent us flowering specimens to be named in the same month of 1873. Quite recently Mr. Green contributed a fine plant of it from Mr. Wilson Saunders' late collection, which came from Paris, with the name I have adopted. I have failed to find any description of this species in any horticultural or botanical work. I may here mention that the difficulty of running down names of Garden plants is, through obvious causes, becoming immense, and will soon be insuperable. I can recommend no more useful object to a Horticultural Society than the organizing a committee for the collection and classification (with references) of the names of all plants introduced into cultivation, together with the countries the plants come from, and their date of introduction.

DESCR. *Stem* probably many feet long, and pendulous from the branches of trees in its native woods, quite glabrous, green, with a faint tinge of brown purple along the margins of the leaf-like articulations, slender and cylindric between the articulations. *Articulations* three to six inches long, by

MARCH 1ST, 1874.

one to one and a half broad, elliptic-lanceolate, acute, narrowed into the petiole-like branches, regularly coarsely obtusely toothed, between coriaceous and fleshy, quite flat, without scales or hairs; midrib and lateral nerves broad and faint, the latter directed to the sinus of the teeth, and unbranched. *Flowers* copiously produced in the axils of the teeth, three-quarters to one inch in diameter, pale straw-coloured, odorous, opening by day. *Ovary* quite naked from a very early stage, sessile, oblong, with four to five obtuse ribs. *Perianth* erecto-patent; leaflets 8–12, narrow-lanceolate, acute or acuminate, the outer rather smaller. *Stamens* numerous, much shorter than the perianth. *Style* slender, stigmas four or five, spreading.—*J. D. H.*

Fig. 1, Ovary style and stigma:—*magnified.*

NOW READY.

Demy 8vo, with 4 double Plates printed in Colours, 8 plain Plates, and five Woodcuts, price 10s. 6d.,

HARVESTING ANTS
AND
TRAP-DOOR SPIDERS.

NOTES AND OBSERVATIONS ON THEIR HABITS AND DWELLINGS.

BY J. TRAHERNE MOGGRIDGE, F.L.S.

L. REEVE & Co., 5, Henrietta Street, Covent Garden.

BOTANICAL PLATES;
OR,
PLANT PORTRAITS.

IN GREAT VARIETY, BEAUTIFULLY COLOURED, 6d. and 1s. EACH.

List of nearly 2000, one stamp.

L. REEVE & Co., 5, Henrietta Street, Covent Garden.

FLORAL PLATES,

BEAUTIFULLY COLOURED BY HAND, 6d. EACH.

A New List of 500 Varieties, one stamp.

L. REEVE & Co., 5, Henrietta Street, Covent Garden.

NOW READY.

BENTHAM AND HOOKER'S
GENERA PLANTARUM.

Part IV. being the first part of Vol. II., comprising Caprifoliaceæ to Compositæ. Price 24s.

L. REEVE & Co., 5, Henrietta Street, Covent Garden.

NOW READY.

Part X., completing the Work, with 10 Coloured Plates, Portrait and Memoir of the Author, and a full Index, price 25s.,

FLORA VITIENSIS,

A DESCRIPTION OF

THE PLANTS OF THE VITI OR FIJI ISLANDS

WITH AN ACCOUNT OF

THEIR HISTORY, USES, AND PROPERTIES.

BY BERTHOLD SEEMANN, PH.D., F.L.S.

The Work complete in One Vol., with 100 Coloured Plates, £8 5s., cloth.

L. REEVE & Co., 5, Henrietta Street, Covent Garden.

Reissue of the Third Series of the Botanical Magazine.

Now ready, Vols. I. and II. price 42s. each (to Subscribers for the entire Series 36s. each).

THE BOTANICAL MAGAZINE, Third Series. By Sir WILLIAM and DR. HOOKER. To be continued monthly.

Subscribers' names received by the Publishers, either for the Monthly Volume or for sets to be delivered complete at 36s. per volume, as soon as ready. The first complete set will be ready in a few days; others will speedily follow.

L. REEVE AND CO., 5, HENRIETTA STREET, COVENT GARDEN.

J. AND W. E. ARCHBUTT

BEG to call attention to their PANTOSCOPIC SPECTACLES with PERISCOPIC LENSES, which obviate all distortion of image. Price 5s. 6d. per pair, with case. POCKET BOTANICAL MAGNIFIERS, high power, from 1s., post free. MICROSCOPES from 12s. 6d. to 12 Guineas.

Catalogues post free.

11, Bridge Street, Westminster, facing the Houses of Parliament.

SECOND-HAND.—THE FLORAL CABINET AND MAGAZINE OF EXOTIC BOTANY, by KNOWLES and WESTCOTT. Illustrated by 137 most beautifully Coloured Plates. Complete 3 vols. half morocco, price £2 2s. Published at £5 5s. unbound.

PLANTÆ ASIATICÆ RARIORES. By N. WALLICH, M. AND PH. D. Parts 2, 3, 4, and 5.

R. J. MITCHELL, 52, PARLIAMENT STREET.

QUADRANT HOUSE,

74, REGENT STREET, AND 7 & 9, AIR STREET, LONDON, W.

AUGUSTUS AHLBORN,

BEGS to inform the Nobility and Gentry that he receives from Paris, twice a week, all the greatest novelties and specialties in Silks, Satins, Velvets, Shawls, &c., and Costumes for morning and evening wear. Also at his establishment can be seen a charming assortment of robes for Brides and Bridesmaids, which, when selected, can be made up in a few hours. Ladies will be highly gratified by inspecting the new fashions of Quadrant House.

From the *Court Journal*:—"Few dresses could compare with the one worn by the Marchioness of Bute at the State Concert at Buckingham Palace. It attracted universal attention, both by the beauty of its texture, and the exquisite taste with which it was designed. The dress consisted of a rich black silk tulle, on which were artistically embroidered groups of wild flowers, forming a most elegant toilette. The taste of the design, and the success with which it was carried out, are to be attributed to the originality and skill of Mr. AUGUSTUS AHLBORN."

Third Series.
No. 352.

VOL. XXX. APRIL. [*Price 3s. 6d. col^d 2s. 6d. plain.*

OR No. 1046 OF THE ENTIRE WORK.

CURTIS'S
BOTANICAL MAGAZINE,

COMPRISING

THE PLANTS OF THE ROYAL GARDENS OF KEW,

AND OF OTHER BOTANICAL ESTABLISHMENTS IN GREAT BRITAIN;
WITH SUITABLE DESCRIPTIONS;

BY

JOSEPH DALTON HOOKER, M.D., C.B., F.R.S., L.S., &c.

Director of the Royal Botanic Gardens of Kew.

Nature and Art to adorn the page combine,
And flowers exotic grace our northern clime.

LONDON:
L. REEVE & CO., 5, HENRIETTA STREET, COVENT GARDEN.
1874.

[*All rights reserved.*]

ROYAL BOTANIC SOCIETY OF LONDON,

GARDENS—REGENT'S PARK.

ARRANGEMENTS FOR 1874.

EXHIBITION OF SPRING FLOWERS, Wednesday, April 22.
SUMMER EXHIBITIONS, Wednesdays, May 20, June 10, and June 24. Gates open at 2 o'clock.
SPECIAL EVENING FETE, Wednesday, July 8. Gates open at 8 o'clock P.M. Evening Dress.
AMERICAN EXHIBITION, Daily, May 25, to June 9.
PROMENADES.—Every Wednesday in May, June, and July, excepting the Exhibition days, commencing May 6. Visitors admitted only by the Special Coloured Orders.
LECTURES in the Museum at 4 o'clock precisely, Fridays, May 15, 22, 29; June 5, 12, 19, 26; July 3.

ROYAL HORTICULTURAL SOCIETY.

MEETINGS AND SHOWS IN 1874.

April	1.	(Fruit and Floral Meeting.)
,,	15.	(Early Rhododendrons.)
May	13.	(Pot Roses.)
,,	27.	(Fruit and Floral Meeting.)
June 4 and 5.		(Great Summer Show.)
,,	17.	(Fruit and Floral Meeting.)
July	1.	(Cut Roses.)
,,	15.	(Zonal Pelargoniums.)
August	5.	(Fruit and Floral Meeting.)
,,	19.	Do.
September	2.	(Dahlias.)
October	7.	(Fruit and Floral Meeting.) (Fungi.)
November	11.	(Fruit and Chrysanthemums.)
December	2.	(Fruit and Floral Meeting.)

"THE GARDEN,"

A Weekly Illustrated Journal devoted solely to Horticulture in all its branches.

THE GARDEN is conducted by WILLIAM ROBINSON, F.L.S., Author of "Hardy Flowers," "Alpine Flowers for English Gardens," "The Parks, Promenades, and Gardens of Paris," &c., and the best Writers in every department of Gardening are contributors to its pages.

The following are some of the subjects regularly treated of in its pages:—

The Flower Garden.	Hardy Flowers.
Landscape Gardening.	Town Gardens.
The Fruit Garden.	The Conservatory.
Garden Structures.	Public Gardens.
Room and Window Gardens.	The Greenhouse and
Notes and Questions.	The Household.
Market Gardening.	The Wild Garden.
Trees and Shrubs.	The Kitchen Garden.

THE GARDEN may be obtained through all Newsagents and at the Railway Bookstalls, at 4d. per copy. It may also be had direct from the Office at 5s. for a Quarter, 9s. 9d. for a Half-year, and 19s. 6d. for a Year, payable in advance; and in Monthly Parts. Specimen Copies (post free) 4½d.

37, Southampton Street, Covent Garden, W.C.

TAB. 6090.

COLCHICUM PARKINSONI.

Native of the Greek Archipelago.

Nat. Ord. MELANTHACEÆ.—Tribe COLCHICEÆ.

Genus COLCHICUM, *Tourn.; (Endl. Gen. Plant.*, p. 137).

COLCHICUM *Parkinsoni;* hysteranthum, cormo magnitudine avellanæ, foliis patulis prostratis et humi appressis, elongato-lanceolatis acuminatis margine insigniter undulatis, perianthii 3-4 poll. diametro tubo albo segmentis patenti-recurvis albis pulcherrime purpureo tessellatis elliptico-lanceolatis subacutis, antheris cœruleis, polline fusco-purpureo, stigmatibus minutis incurvis perianthii segmentis multo brevioribus.

? COLCHICUM chionense, *Haw.; ex Kunth. Enum.*, vol. iv. p. 139, *sub* C. variegatum.

C. Fritillaricum Chiense, *Parkins. Parad.*, p. 155, f. 5, et p. 156.

This charming Meadow Saffron appears to have been actually lost sight of by botanists for nearly two and a half centuries. It is originally very accurately described and rudely figured by Parkinson, in the "Paradisus Terrestris," published in 1629, where it is distinguished from the other tessellated-flowered Colchicums by its smaller size, brighter, clearer colouring, and the undulated leaves lying flat on and appressed to the ground.

Ray, in his "Historia Plantarum," p. 1172, published in 1688, keeps up Parkinson's plant under his name, but adds to it Cornutis's *C. variegatum* as the same thing; in this he was mistaken, for a reference to Cornutis's work, published in 1635, with a rude woodcut, proves that his is a very different plant, a native of Messina, and is probably that now known as *C. Bivonæ*, Guss. The plant now called *variegatum*, and which is supposed to be the Linnæan one, is also a native of Greece, and is figured at Tab. 1028 of this work (copied and reversed in Reichenbach's "Flora Exoticæ," t. 57, without acknowledgment). This, Mr. Baker informs me, is a much larger species than the subject of the present plate, with less pronounced and coarser tessellation, and having suberect leaves

APRIL 1ST, 1874.

a foot high and less undulated. It is known under the name of *C. variegatum, tessellatum,* and *agrippinum* in English gardens, and is liable to be killed in severe winters.

With regard to Haworth's name of *chionense*, cited without a reference by Kunth, I can nowhere else find it; and having no means of knowing to what plant he applied it, I hesitate to apply it to this, which should henceforth bear the name of the acute old botanist who first published it, and whose quaint and characteristic description I here give at length :—

"This most beautiful Saffron flower riseth up with his flowers in the Autumn, as the others before specified do, although not of so large a size, yet far more pleasant and delightful in the thick, deep blew or purple-coloured beautiful spots therein, which make it excel all others whatsoever: the leaves rise up in the Spring, being smaller then the former, for the most part three in number, and of a paler or fresher green colour, lying close upon the ground, broad at the bottom, a little pointed at the end, and twining or folding themselves in and out at the edges, as if they were indented. I have not seen any seed it hath born: the root is like unto the others of this kinde, but small and long, and not so great: it flowreth later for the most part then any of the other, even not until November, and is very hard to be preserved with us, in that for the most part the root waxeth lesse and lesse every year, our cold country being so contrary unto his natural, that it will scarce shew his flower; yet when it flowreth any thing earlie, that it may have any comfort of a warm Sun, it is the glory of all these kindes."— *Paradisus Terrestris,* p. 156.—*J. D. H.*

TAB. 6091.

BESCHORNERIA TONELII.

Native of Mexico.

Nat. Ord. AMARYLLIDEÆ.—Tribe AGAVEÆ.

Genus BESCHORNERIA, *Kunth*; *(Kunth, Enum. Plant.*, vol. v. p. 844).

BESCHORNERIA *Tonelii*; foliis recurvis pedalibus elliptico-lanceolatis acuminatis asperulis subtillissime denticulatis, scapo 4-pedali sanguineo-purpureo, panicula 2-pedali inclinata, ramis paucis gracilibus patentibus sparsifloris, bracteis ovato-lanceolatis acuminatis pallidis, floribus nutantibus et pendulis, pedicellis ovariis perianthiique segmentis basi et dorso late sanguineo-purpureis, perianthii segmentis acutis viridibus.

BESCHORNERIA Tonelii, *Jacobi in Otto Hamburg. Garten- und Blumenz.*, vol. xx. p. 503, *sine descriptione*.

Had I not examined this plant in a living state in Mr. Wilson Saunders's garden, where it flowered in May of last year, I should certainly have identified it with the original *B. tubiflora*, Kunth, as figured at Tab. 4642 of this work; nor am I now sure, after a comparison of these real or supposed species, that they are more than varieties of one. The present is of a laxer habit, has much broader leaves, and brighter red purple scape and panicle, the latter with drooping branches; it has also rather longer more pendulous flowers with more acute perianth-segments. In all other respects, and especially in the floral organs, the two supposed species appear to be identical.

According to General Jacobi, who (in Otto's work, cited above), has given a sketch of the genera and species of *Agaveæ*, the genus *Beschorneria* contains four species, of which two are now figured in this Magazine, and the others, *B. yuccoides* and *B. Parmentieri* (*Yucca Parmentieri*, Roezl), are unknown to me. Unfortunately General Jacobi gives no description of *B. Tonelii*, his conspectus of *Aloineæ*, which was commenced in the work referred to, not having been continued to *Beschorneria*, and I am therefore dependent on the authority of Mr. Wilson Saunders's garden for the name this plant bears.

DESCR. *Stem* very short. *Leaves* few, spreading, fifteen to twenty inches long, by two and a half inches broad, acuminate and keeled beneath towards the tip, minutely serrulate, scaberulous above, very glaucous, thick and hard, contracted into a flat thick petiole an inch broad. *Scape* four feet high, as thick as the middle finger below, and as well as the inflorescence of a bright red-purple colour. *Panicle* two feet long, slender, inclined, with few lax spreading simple branches bearing distant fascicles of two to five flowers; bracts several to each fascicle of flowers, three quarters to one and a quarter inches long, ovate-lanceolate, acuminate, spreading, membranous, pale. *Flowers* two and a half inches long, drooping, on slender pedicels half to three-quarters inch long. *Ovary* one inch long, obtusely 3-gonous, dark red-purple, 6-grooved. *Perianth* tubular; segments linear, slightly dilated at the rather spreading acute tip, dark blood-red below and on the midrib, the rest very bright verdigris green. *Stamens* nearly as long as the perianth, filaments dilated above the base; anthers linear-oblong. *Style* rather longer than the stamens, base conical; stigma obscurely 3-lobed.—*J. D. H.*

Fig. 1, Top of ovary with stamens and style:—*magnified*.

TAB. 6092.

ACONITUM HETEROPHYLLUM.

Native of the Western Himalaya.

Nat. Ord. RANUNCULACEÆ.—Tribe HELLEBOREÆ.

Genus ACONITUM, *Linn.*; (*Benth. & Hook. f. Gen. Plant*, vol. i. p. 9).

ACONITUM *heterophyllum;* caule erecto robusto simplici v. ramoso glabro superne pubescente, foliis radicalibus petiolatis rotundato-reniformibus v. cordatis obscure 5-lobis grosse duplicatim inciso-dentatis v. lobulatis glabris, caulinis late cordatis sessilibus brevissime petiolatis v. amplexicaulibus, racemis lateralibus v. terminalibus multifloris dense v. laxifloris, pedicellis erectis, bracteolis 2-3, floribus cœruleis v. ochroleucis purpureo cœruleo ve venosis, sepalis puberulis, supremo valde convexo, lateralibus oblique ovatis, antico lanceolato sinuoso, petalis ungue late lineari subincurvo apice subgloboso inflato ecalcarato, carpellis 5 pubescentibus, folliculis erectis.

ACONITUM heterophyllum, *Wall. Cat.* 4722; *Royle Ill. Pl. Himal.*, p. 56, t. 13; *Hook. f. & Thoms. Flor. Ind.*, vol. i. p. 58; *Hook. f. Flor. Brit. Ind.*, vol. i. p. 29.

A. cordatum, *Royle Ill.* p. 56.

A. Atees, *Royle in Journ. As. Soc. Beng.*, vol. i. p. 459.

The subject of the present plate is a very interesting plant, as being, though a member of a most poisonous genus, in extensive use as a tonic medicine throughout N. India, under the name of Atees or Atis. It inhabits the whole Western Himalaya, from Kumaon to Kashmir, at elevations from 8–13,000 ft., growing in moist places, at the edge of forests, &c. It is a near ally of the famous Bikh poison of the same mountains, which does not seem to differ from our deadly *A. Napellus* (Monkshood). For the specimen here figured I am indebted to Colonel G. Smyth of Wetten-le-Wold, Louth, who cultivated it in his garden from Himalayan seed, and communicated it to Kew in August of last year.

Dr. Royle says of this species:—" In the native works on Materia Medica, as well as in the common Persian and Hindoostanee and English Dictionaries, *Atees* is described as being the root of an Indian plant used in medicine. This, the author learnt, was the produce of the Himalayas; he therefore sent to one of the commercial entrepôts situated at the foot of the hills, and procured some of the root, making inquiries respecting the part of the mountains whence it was procured. The plant-collectors, in their next excursion, were directed to bring the plant, with the root attached to it, as the only evidence which would be admitted as satisfactory. The first specimens thus procured are represented in Plate 13, and the root *Atees* having been thus ascertained to be the produce of a new species of Aconite, was named *Aconitum Atees* (Journ. Asiat. Soc., i. p. 459); but which has since been ascertained to be the *Aconitum heterophyllum* of Dr. Wallich. The roots obtained in different parts of the country resemble one another, as well as those attached to the plant. They are about an inch in length, of an oblong oval-pointed form, light-greyish colour externally, white in the inside, and of a pure bitter taste. That its substance is not so injurious as the *Bish*, I conclude from its being attacked by insects, while the other remains sound and untouched. The natives describe it as being of two kinds, one black, the other white, and both as bitter, astringent, pungent, and heating, aiding digestion, useful as a tonic, and aphrodisiac."—Royle Ill. Pl. Himal., p. 48.

Fig. 1, Pedicel with bract, stamen, and two lateral sepals :—*magnified*.

TAB. 6093.

PANAX SAMBUCIFOLIUS.

Native of New South Wales and Victoria.

Nat. Ord. ARALIACEÆ.—Series PANACEÆ.

Genus PANAX, *Linn.*; (*Benth. & Hook. f. Gen. Plant.*, vol. i. p. 938).

PANAX *sambucifolius;* glaberrimus, foliis pinnatis 2-pinnatisque, foliolis polymorphis sessilibus petiolulatisve ellipticis v. lanceolatis integerrimis dentatis lobulatis v. pinnatifidis subtus glaucis, rachi interdum dilatata ad nodos articulata, umbellis terminalibus et axillaribus corymbosis paniculatis v. racemosis, calycis limbo brevissimo sinuato 4-5-dentato, fructu baccato globoso aquoso translucido, pyrenis plano-convexis dorso obtuse costatis.

PANAX sambucifolius, *Sieb. in DC. Prod.*, vol. iii. p. 255; *Benth. Fl. Austral*, vol. iii. p. 382.

P. angustifolius *et* P. dendroides, *F. Muell. in Trans. Phil. Inst. Vict.*, vol. i. p. 42; *Plant. Vict.*, t. 28.

NOTHOPANAX sambucifolius, *Seem. Flor. Viti.*, p. 115.

The singular beauty of the translucent berries which persist for a long time on the plant, recommend the latter for cultivation. These resemble white currants in form and transparency, but have a faint blue tinge, and each is capped by a minute black calyx-limb, and two thread-like diverging or recurved styles. It is a native of extra-tropical Eastern Australia, extending from north of the New South Colony to Victoria; and a very similar plant (of which I have seen the leaves only) has been sent from Tasmania. Like so many *Araliaceæ*, the Ivy notably, the leaf varies most extraordinarily, being simply or doubly pinnate, and the leaflets being quite entire, toothed, lobed, or pinnatifid, and the petiole flat or dilated between the leaflets. The flowers are small and insignificant; they appear in spring, and the beautiful berries ripen in September.

Panax sambucifolius was introduced into Kew from the Melbourne Botanic Garden by Baron Mueller, and flowered for the first time in 1873.

APRIL 1ST, 1874.

DESCR. A shrub or small tree, everywhere quite glabrous; branches slender, green. *Leaves* three to ten inches long, pinnate or 2-pinnate; leaflets one to three inches long, sessile or petioled, elliptic or lanceolate, quite entire, toothed, lobed or pinnatifid, and cut into distant short or long, broad or narrow lobes, glaucous beneath; rachis simple winged or dilated and leafy. *Umbels* small, half an inch in diameter, green, in corymbs, racemes, or panicles; peduncles slender; pedicels very short, jointed below the flower. *Flowers* one-sixth of an inch in diameter. *Calyx-tube* hemispheric, contracted into a short stipe; limb 4–5-toothed. *Petals* four to five, spreading with incurved tips; those of the female flowers larger and often cohering at the tips, smaller and more spreading in the males. *Fruit* a watery transparent berry, one-third of an inch in diameter, crowned with a minute black calyx limb, and two slender recurved styles; pyrenes two, plano-convex, with two striate dorsal ribs.—*J. D. H.*

Fig. 1 and 2, Male flowers; 3, calyx, and styles of ditto; 4, ripe fruit; 5, transverse section of ditto:—*all magnified.*

TAB. 6094.

EPIDENDRUM CRINIFERUM.

Native of Costa Rica.

Nat. Ord. ORCHIDEÆ.—Tribe EPIDENDREÆ.

Genus EPIDENDRUM, *Linn.*; (*Lindl. Fol. Orchid. Epidendrum*).

EPIDENDRUM (Spathium) *criniferum*; caulibus repentibus et fasciculatis simplicibus gracilibus ima basi tuberosis superne foliosis, foliis subdistichis, vaginis teretibus, lamina sessili lineari-lanceolata acuta patenti-recurva, spathis pluribus viridibus lineari-oblongis apice truncatis, racemo terminali subsessili, bracteis minutis appressis, floribus 2 poll. diam., sepalis subulato-lanceolatis acuminatis aureis rubro-fusco maculatis, petalis æquilongis fere filiformibus, labelli laciniis lateralibus semi-ovatis incurvis lateribus in processus subulatos undulatos fissis, lacinia media porrecta lineari angusta, disco callis 2 parvis supra stigma ornato.

EPIDENDRUM criniferum, *Reichb. f. in Gard. Chron.* 1871, p. 1291.

This belongs to a large West-Indian and South-American section of *Epidendrum*, of which many species have been described by Lindley in his "Folia Orchidacea," by Reichenbach, and by others; of the latter the Cuban *E. rivulare*, Lindl., according to Reichenbach, comes nearest to this, but differs in the longer and narrower leaves, shorter bristles on the lip, and different midlobe of the latter. Reichenbach further observes that the lip of this is totally white, but in the specimen here figured it is blotched with pale red, and the midlobe is yellow.

Epidendrum criniferum was sent for figuring by Messrs. Veitch, in whose splendid collection of Orchids it flowered in January of the present year.

DESCR. *Stems* a foot high, in tufts from a creeping rootstock, slender, green, leafy, as stout as a swan's-quill, the basal joints swollen and half an inch in diameter. *Leafsheathes* one inch long, green, cylindric, with a truncate mouth; leaf-blade, three to four inches long by half inch broad, spreading and recurved, linear-lanceolate, acute, sessile,

APRIL 1ST, 1874.

dark-green above, pale beneath, margins subrecurved, midrib keeled. *Spathes* about three, terminal, erect, an inch long, linear-oblong, truncate. *Racemes* terminal, subsessile, inclined, about 6-flowered; rachis green, one to one and a half inches long; bracts small, green, appressed to the slender pedicel. *Ovary* slender, with the pedicel one and a half inches long. *Perianth* nearly two inches across. *Sepals* equal, flat, spreading, subulate-lanceolate, acuminate, golden-yellow blotched with chestnut-brown. *Petals* as long and similarly coloured, but extremely narrow. *Lip* adnate to the column, 3-lobed, lateral lobes semi-ovate with crinite margins, the setæ curved upwards and waved; disk white with very pale pink blotches and two tubercular calli close under the stigma; midlobe straight, yellow, very narrow, acute, extending as far as the petals. *Column* white.—*J. D. H.*

Fig. 1, Ovary, column, and lip:—*magnified*.

Tab. 6095.

RHOPALA Pohlii.

Native of Brazil.

Nat. Ord. PROTEACEÆ.—Tribe GREVILLEÆ.

Genus RHOPALA, *Schreb.; (Meissner in DC. Prod.,* vol. xiv. p. 424).

RHOPALA *Pohlii;* arborescens, ramulis petiolisque fusco-tomentosis demum glabratis, foliis imparipinnatis, foliolis petiolatis oblique ovatis ellipticis-ovatisve acuminatis grosse serratis supra glabris nitidis subtus costatis reticulatis laxe tomentosis v. glabratis, racemis erectis simplicibus v. compositis solitariis sessilibus aureo-lanatis, pedicellis bracteis parvis duplo longioribus, perianthio clavato ad medium 4-fido, segmentis anguste spathulatis, ovario villoso, stylo gracili.

RHOPALA Pohlii, *Meissn. in Mart. Fl. Bras.,* fasc. xiv. p. 89, t. 33; *et in DC. Prod.,* vol. xiv. t. 433.

R. corcovadensis, *Hort.*

The genus *Rhopala* is one of the few American representatives of the Old World *Proteaceæ*, and is confined to the tropical and south temperate regions of the New World, where nearly forty species have been found, many of them in Brazil. They are, for the most part, exceedingly handsome evergreen plants, with dark-green shining coriaceous leaves, and insignificant blossoms, usually dotted with a rusty or golden pubescence. The present species is a native of the province of Minas Geraes, in Brazil, and of the neighbourhood of Rio de Janeiro, whence it was introduced into Kew many years ago, from a Belgian garden I believe, probably Mr. Linden's, under the name of *R. corcovadensis.* It has flowered repeatedly in the palm-house early in the year.

DESCR. A tree. *Branches* clothed with dense, bright, rusty-coloured woolly tomentum. *Leaves* a foot long and upwards, arched, pinnate, with five to eight pairs of subopposite and alternate pinnules, rachis, petiole, and petiolules villous; pinnules three to five inches long, on stout petiolules, which are sometimes an inch long, obliquely ovate or elliptic-ovate,

APRIL 1ST, 1874.

acuminate, coarsely acutely serrate, bright-green above, paler and tomentose but at length glabrate beneath, and reticulated with prominent veins. *Racemes* axillary, erect, three to five inches long, subsessile, narrow, simple or compound at the base, covered with orange-red velvety pubescence; pedicels strict, erect, much larger than the small bracts. *Flowers* one-third of an inch long. *Perianth* clavate before opening, divided to below the middle into four narrowly spathulate segments. *Ovary* narrow-ovoid, villous; style slender, stigma clavate.—*J. D. H.*

Fig. 1, Flower; 2, pedicel and pistil:—*magnified.*

Now Ready, Part II., with 4 Coloured Plates, Royal 4to, price 5s.

ORCHIDS,

AND

How to Grow them in India & other Tropical Climates.

BY

SAMUEL JENNINGS, F.L.S., F.R.H.S.,

Late Vice-President of the Agri-Horticultural Society of India.

NOW READY, Part II., 10s. 6d.

FLORA OF INDIA.

BY

DR. HOOKER, C.B., F.R.S.,

Assisted by various Botanists.

NOW READY, Vol. VI., 20s.

FLORA AUSTRALIENSIS.

A Description of the Plants of the Australian Territory. By GEORGE BENTHAM, F.R.S., assisted by BARON FERDINAND MUELLER, C.M.G., F.R.S. Vol. VI. Thymeleæ to Dioscorideæ.

NOW READY.

LAHORE TO YARKAND.

Incidents of the Route and Natural History of the Countries traversed by the Expedition of 1870, under T. D. FORSYTH, Esq., C.B. By GEORGE HENDERSON, M.D., F.L.S., F.R.G.S., Medical Officer of the Expedition, and ALLAN O. HUME, Esq., C.B., F.Z.S., Secretary to the Government of India. With 32 Coloured Plates of Birds and 6 of Plants, 26 Photographic Views of the Country, a Map of the Route, and Woodcuts. Price 42s.

In the Press and shortly to be published, in one large Volume, Royal 8vo, with numerous Coloured Plates of Natural History, Views, Map and Sections. Price 42s.

To Subscribers forwarding their Names to the Publishers before publication, 36s.

ST. HELENA:

A

Physical, Historical, and Topographical Description of the Island.

INCLUDING ITS

GEOLOGY, FAUNA, FLORA, AND METEOROLOGY.

BY

JOHN CHARLES MELLISS, C.E., F.G.S., F.L.S.

LATE COMMISSIONER OF CROWN PROPERTY, SURVEYOR AND ENGINEER OF THE COLONY.

L. REEVE & CO., 5, HENRIETTA STREET, COVENT GARDEN.

RE-ISSUE of the THIRD SERIES of the BOTANICAL MAGAZINE.

Now ready, Vols. I. to IV., price 42s. each (to Subscribers for the entire Series 36s. each).

THE BOTANICAL MAGAZINE, Third Series. By Sir WILLIAM and DR. HOOKER. To be continued monthly.

Subscribers' names received by the Publishers, either for the Monthly Volume or for sets to be delivered complete at 36s. per volume, as soon as ready. The first complete set will be ready in a few days; others will speedily follow.

L. REEVE AND CO., 5, HENRIETTA STREET, COVENT GARDEN.

J. AND W. E. ARCHBUTT

BEG to call attention to their PANTOSCOPIC SPECTACLES with PERISCOPIC LENSES, which obviate all distortion of image. Price 5s. 6d. per pair, with case. POCKET BOTANICAL MAGNIFIERS, high power, from 1s., post free. MICROSCOPES from 12s. 6d. to 12 Guineas.

Catalogues post free.

11, Bridge Street, Westminster, facing the Houses of Parliament.

SECOND-HAND.—THE FLORAL CABINET AND MAGAZINE OF EXOTIC BOTANY, by KNOWLES and WESTCOTT. Illustrated by 187 most beautifully Coloured Plates. Complete 3 vols. half morocco, price £2 2s. Published at £5 5s. unbound.

PLANTÆ ASIATICÆ RARIORES. By N. WALLICH, M. AND PH. D. Parts 2, 3, 4, and 5.

R. J. MITCHELL, 52, PARLIAMENT STREET.

QUADRANT HOUSE,
74, REGENT STREET, AND 7 & 9, AIR STREET, LONDON, W.

AUGUSTUS AHLBORN,

BEGS to inform the Nobility and Gentry that he receives from Paris, twice a week, all the greatest novelties and specialties in Silks, Satins, Velvets, Shawls, &c., and Costumes for morning and evening wear. Also at his establishment can be seen a charming assortment of robes for Brides and Bridesmaids, which, when selected, can be made up in a few hours. Ladies will be highly gratified by inspecting the new fashions of Quadrant House.

From the *Court Journal*:—"Few dresses could compare with the one worn by the Marchioness of Bute at the State Concert at Buckingham Palace. It attracted universal attention, both by the beauty of its texture, and the exquisite taste with which it was designed. The dress consisted of a rich black silk tulle, on which were artistically embroidered groups of wild flowers, forming a most elegant toilette. The taste of the design, and the success with which it was carried out, are to be attributed to the originality and skill of Mr. AUGUSTUS AHLBORN."

Third Series.

No. 353.

VOL. XXX. MAY. [*Price* 3s. 6d. col^{d.} 2s. 6d. *plain.*

OR No. 1047 OF THE ENTIRE WORK.

CURTIS'S
BOTANICAL MAGAZINE,

COMPRISING

THE PLANTS OF THE ROYAL GARDENS OF KEW,

AND OF OTHER BOTANICAL ESTABLISHMENTS IN GREAT BRITAIN,
WITH SUITABLE DESCRIPTIONS;

BY

JOSEPH DALTON HOOKER, M.D., C.B., F.R.S., L.S., &c.

Director of the Royal Botanic Gardens of Kew.

Nature and Art to adorn the page combine,
And flowers exotic grace our northern clime.

LONDON:
L. REEVE & CO., 5, HENRIETTA STREET, COVENT GARDEN.
1874.

[*All rights reserved.*]

ROYAL BOTANIC SOCIETY OF LONDON,

GARDENS—REGENT'S PARK.

ARRANGEMENTS FOR 1874.

SUMMER EXHIBITIONS, Wednesdays, May 20, June 10, and June 24. Gates open at 2 o'clock.

SPECIAL EVENING FETE, Wednesday, July 8. Gates open at 8 o'clock P.M. Evening Dress.

AMERICAN EXHIBITION, Daily, May 25, to June 9.

PROMENADES.—Every Wednesday in May, June, and July, excepting the Exhibition days, commencing May 6. Visitors admitted only by the Special Coloured Orders.

LECTURES in the Museum at 4 o'clock precisely, Fridays, May 15, 22, 29; June 5, 12, 19, 26; July 3.

ROYAL HORTICULTURAL SOCIETY.

MEETINGS AND SHOWS IN 1874.

May	13.	(Pot Roses.)
,,	27.	(Fruit and Floral Meeting.)
June 4 and 5.		(Great Summer Show.)
,,	17.	(Fruit and Floral Meeting.)
July	1.	(Cut Roses.)
,,	15.	(Zonal Pelargoniums.)
August	5.	(Fruit and Floral Meeting.)
,,	19.	Do.
September	2.	(Dahlias.)
October	7.	{ (Fruit and Floral Meeting.) (Fungi.)
November 11.		(Fruit and Chrysanthemums.)
December 2.		(Fruit and Floral Meeting.)

"THE GARDEN,"

A Weekly Illustrated Journal devoted solely to Horticulture in all its branches.

THE GARDEN is conducted by WILLIAM ROBINSON, F.L.S., Author of "Hardy Flowers," "Alpine Flowers for English Gardens," "The Parks, Promenades, and Gardens of Paris," &c., and the best Writers in every department of Gardening are contributors to its pages.

The following are some of the subjects regularly treated of in its pages:—

The Flower Garden.	Hardy Flowers.
Landscape Gardening.	Town Gardens.
The Fruit Garden.	The Conservatory.
Garden Structures.	Public Gardens.
Room and Window Gardens.	The Greenhouse and
Notes and Questions.	The Household.
Market Gardening.	The Wild Garden.
Trees and Shrubs.	The Kitchen Garden.

THE GARDEN may be obtained through all Newsagents and at the Railway Bookstalls, at 4d. per copy. It may also be had direct from the Office at 5s. for a Quarter, 9s. 9d. for a Half-year, and 19s. 6d. for a Year, payable in advance; and in Monthly Parts. Specimen Copies (post free) 4½d.

37, Southampton Street, Covent Garden, W.C.

Tab. 6096.

XIPHION Sisyrinchium.

Native of the Mediterranean Region.

Nat. Ord. Iridaceæ.—Tribe Irideæ.

Genus Xiphion, *Tourn.*; (*Tab. nost.* 5890).

Xiphion *Sisyrinchium;* bulbo globoso fibroso-tunicato, caule gracili sæpius flexuoso 2–6-floro, foliis sub-binis elongato-subulatis falcatis dorso semiteretibus, spathis 1½-pollicaribus ventricosis, ovario subsessili, perianthii tubo gracili subunciali, limbo tenuissimo cœruleo-violaceo 1–1½ poll. diametro, segmentis exterioribus obovato-spathulatis, interioribus angustis paulo brevioribus erectis, filamentis deorsum stylo adnatis, stigmatibus segmentis perianthii interioribus æquilongis profunde 2-fidis, lobis erectis parallelis subulato-lanceolatis.

X. Sisyrinchium, *Baker in Seemann Journ. Bot.*, vol. ix. p. 42.

Iris Sisyrinchium, *Linn. Sp. Pl.*, vol. i. p. 59; *Sibth. Fl. Græc.*, vol. i. p. 30, t. 42; *Cav. Ic.*, t. 193; *Rédouté Lil.*, t. 29 et 458.

I. Ægyptia, *Delile Fragm. Fl. Arab.*, p. 6.

I. fugax, *Ten. Fl. Neap.*, vol. i. p. 15, t. 4.

Gynandriris Sisyrinchium, *Parl. Nuov. Gen.*, p. 52, *Flor. Ital.*, vol. iii. p. 309; *Godr. et Gren. Fl. France*, vol. iii. p. 246; *Klatt in Linnæa*, vol. xxxiv. p. 577.

Moræa Sisyrinchium, *Gawl. Bot. Mag.*, t. 1407.

M. fugax, *Ten. Syll.*, p. 26.

M. Tenoreana, *Sweet. Brit. Fl. Gard.*, t. 110.

This lovely little plant is the most widely diffused of all the *Irideæ*, extending from Spain and Marocco to Turkey and Egypt in Europe and Africa respectively, and thence eastward through Syria and Arabia to Affghanistan and Beloochistan. It further, probably, passes the confines of the British Indies, as my correspondent, Dr. Aitcheson, informs me by a letter just received, that he has found a bulbous *Iris* in the North-Western Punjab, which, from his description, may well be this.

It will be observed that this is the *Moræa Sisyrinchium* figured in this work (Tab. 1407), but so indifferently as

May 1st, 1874.

hardly to be recognisable. It was introduced into England before the days of Gerard (1597), but is still scarce, being often killed by frost. The plants here figured flowered at Kew in May of last year, from bulbs sent by D. Hanbury, Esq., F.R.S., from Calabria. The bulbs are said (*Bot. Mag. l. c.*) to be eaten in Spain and Portugal, whence Gerard and Parkinson called them Spanish nuts; but I cannot confirm this statement.

DESCR. *Bulb* the size of a large hazel-nut, globose, clothed with matted fibres. *Stem* six to twelve inches high, stout or flexuous, 2-6-flowered. *Leaves* dark green, usually twin, distichous, spreading or recurved, elongate-subulate very concave, rounded at the backs, about as long as the stem. *Spathes* ventricose, membranous acuminate. *Flowers* one and a half inches in diameter, blue-purple, with a yellow oblong spot on the disk of the outer perianth segments. *Ovary* slender, about one inch long. *Perianth-tube* as long and more slender; outer segments of limb reflexed, oblong-spathulate, obtuse; inner narrow-lanceolate, erect, much paler. *Stigmas* erect, as long as the inner segments, bifid to the middle, segments subulate, parallel, erect.—*J. D. H.*

Fig. 1, Apex of stigma:—*magnified.*

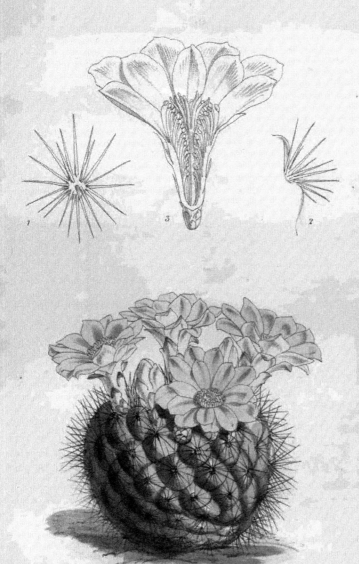

Tab. 6097.

ECHINOCACTUS Cummingii.

Native of Bolivia.

Nat. Ord. CACTEÆ.—Tribe ECHINOCACTEÆ.

Genus ECHINOCACTUS, *Link & Otto* ; (*Benth. & Hook. f. Gen. Pl.*, vol. i. p. 848).

ECHINOCACTUS *Cummingii;* subglobosus, griseo-virescens, tuberculis ⅓-½ poll. diam. distinctis subhemisphericis spiraliter dispositis centro depressis, areolis parvis fere circularibus, demum nudis, spinis exterioribus 15–20 patentibus gracilibus strictis ad ¼ poll. longis pallide flavescentibus superioribus longioribus centralibus fortioribus, floribus 1 poll. diametro aureis, perianthii tubo infundibuliformi, laciniis ad 40 aureis extimis tubum efformantibus brevibus imbricatis apicibus sanguineis, intimis numerosis patentibus lineari-oblongis obtusis, staminibus confertis auries tubo fere immersis, antheris parvis, stylo columnari, stigmatibus 7-8 erectis cylindraceis.

ECHINOCACTUS Cummingii, *Salm-Dyck. Cact. Hort. Dyck. Cult.* p. 174; *Labourét Monog. Cact.*, p. 264.

A very elegant little globose Cactus, with rather large bright golden flowers, communicated to Kew by Mr. Pferfsdorff in June of last year: it is stated by Labourét and Salm-Dyck to be a native of Bolivia, and to be very rare in Europe, but one specimen according to the former author existing in France (in 1847), which was in the collection of M. Andry, of Chaillot. I give it the name under which Mr. Pferfsdorff sends it, assuming it to be correct; it agrees with Labourét's character in everything but the size of the flowers, which are described as "petites," whereas these are of considerable size in proportion to the size of the plant.

DESCR. *Stem* in our specimen two and a half inches in diameter, nearly globose, of a grey-green colour, hardly shining, contracted slightly at the base. *Tubercles* about one third of an inch in diameter, arranged in spirals, subhemispherical, base obtusely quadrangular, with a depression at the top in which the areole is placed. *Areole* small, nearly circular, outer spines about fifteen to twenty, strict, slender, erecto-patent, pale yellowish, the upper rather the longest,

MAY 1ST, 1874.

central two or three shorter and stouter. *Flowers* numerous, sessile. *Perianth* golden yellow, one inch in diameter, and about as long; tube funnel-shaped, clothed with lax imbricating oblong scales tipped with red (the outer segments); inner segments numerous, spreading, linear-oblong, obtuse, flat and overlapping. *Stamens* lining the whole perianth-tube, the innermost much the shortest, filaments strict slender, anthers minute, yellow. *Style* rather stout with seven to eight erect, thickly filiform stigmas.—*J. D. H.*

Fig. 1, Areole and spines; 2, apex of tubercle and spines; 3, vertical section of flower:—*all magnified.*

Tab. 6098.

EPIDENDRUM (Barkeria) LINDLEYANUM.

Native of Costa Rica.

Nat. Ord. Orchideæ.—Tribe Epidendreæ.

Genus Epidendrum, *L.*; (*Lindley Fol. Orchid. Epidendrum*).

Epidendrum (Barkeria) *Lindleyanum*; caule tereti ramoso non bulboso, foliis elliptico-oblongis acutis patenti-recurvis planiusculis enerviis subtus carinatis, vaginis breviusculis, scapo gracili terminali laxi-plurifloro, bracteis viridibus subulato-lanceolatis, floribus patentibus purpureis, pedicello ovarioque gracillimis, sepalis elliptico-lanceolatis acuminatis, petalis consimilibus sed latioribus, labello oblongo-quadrato apiculato ungue brevi basi columnæ adnato, disco albo 2-carinato, columna clavata anguste alata, apice 3-loba.

E. Lindleyanum, *Reichb. f. in Walp. Ann.*, vol. vi. p. 375.

Barkeria Lindleyana, *Batem. in Bot. Reg.*, vol. xxviii. *Misc.*, p. 2, *et in Orchid. Mex.*, tab. xxviii.; *Paxt. Mag. Bot.*, vol. xiii. p. 193.

According to Reichenbach the genus *Barkeria* falls into *Epidendrum*, and no doubt rightly; and I cannot but wonder how it was that Lindley, when revising the latter genus in his "Folia Orchidacea," failed to perceive that *Barkeria* must be united with it, the character of the winged column being a very trivial and far from conspicuous one, and the amount of adnation between the column and lip in *Epidendrum* itself being, as in *Barkeria*, very variable.

Reichenbach further places all the Barkerias in a natural section of *Epidendrum*, which he designates as *Amblostoma*, characterized by a free or nearly free lip; to which section the name of *Barkeria*, now so well known amongst horticulturists, may be equally well applied.

E. Lindleyanum is a native of Costa Rica, where it was discovered by the late Mr. Skinner. The specimen here figured flowered in Mr. Veitch's establishment in December last, and has larger flowers, of a paler colour than those of the plant figured by Bateman, and by Paxton in his Magazine.

Descr. *Stems* a foot or more high, terete, sparingly

branched and rooting at the nodes, clothed below with short pale sheaths, and above with subdistichous leaves, as thick as a goosequill. *Leaves* four to five inches long, spreading and recurved, elliptic-lanceolate, acute, nerveless, keeled at the back. *Scape* terminal, very slender, inclined, many-flowered. *Flowers* lax, two inches in diameter, purple except the white disk of the lip. *Bracts* green, slender, longer or shorter than the slender pedicel, but never exceeding the ovary, which is also slender. *Sepals* spreading and recurved, elliptic-lanceolate, acuminate. *Petals* similar, but broader. *Lip* oblong-quadrate, apiculate, flat with subserrate edges; claw adnate to the lower one-third of the column; disk white with two keels. *Column* clavate, narrowly winged, purple, three-lobed at the top.—*J. D. H.*

Fig. 1, Column :—*magnified.*

TAB. 6099.

SENECIO (KLEINIA) ANTEUPHORBIUM.

Native of South Africa.

Nat. Ord. COMPOSITÆ.—Tribe SENECIONIDEÆ.

Genus SENECIO, *L.*; (*Benth. and Hook. f. Gen. Pl.*, vol. ii. p. 446).

SENECIO (Kleinia) *Anteuphorbium;* glaberrimus, carnosus, erectus, ramosus, caule crasso cylindraceo ad nodos constricto, foliis pollicaribus sparsis erectis oblongis lineari-oblongisve obtusis acutisve integerrimis carnosis, petiolo brevissimo secus caulem linea triplici deducto, capitulis pollicaribus crasse pedunculatis solitariis erectis, bracteis sparsis anguste linearibus, involucri squamis anguste linearibus acutis numerosis, floribus omnibus tubulosis flavis, pappi setis tenuissimis, acheniis lævibus, styli ramis apice acutis.

KLEINIA Anteuphorbium, *DC. Prodr.*, vol. vi. p. 338; *Harv. et Sond. Fl. Cap.*, vol. iii. p. 319.

CACALIA Anteuphorbium, *Linn. Sp. Pl.*, p. 1168; *Willd. Sp. Pl.*, vol. iii. p. 1725; *Ait. Hort. Kew.*, vol. iv. p. 497.

ANTEUPHORBIUM *Bauh. Pinax*, 387; *Dod. Pempt.*, 3, lib. ii. p. 378; *Lob. Ic.*, vol. ii. p. 26; *Moris Hist.*, vol. iii. p. 345; *Dill. Hort. Elth.*, p. 63, t. 55, f. 2, 3.

The subject of the present plate is one of the oldest Cape plants in cultivation, having, according to Dodonæus, been brought to Europe in 1570, and cultivated in England in Gerard's garden in 1596. Nevertheless, its recent South African habitat is up to this date unknown, no accurate description of it has hitherto appeared, and it has been but once seen in flower in Europe, until I received the specimen from which the accompanying drawing was made in January last from Mr. Thomas Hanbury's garden at Palazzo Orengo, near Mentone. Dillenius indeed, so early as 1732, points out the rarity of its flowering, adding that, as his work "The Hortus Elthamensis," was passing through the press, he received a flowering specimen from a Mr. Powers, gardener to Mr. Blaithwaits, at Dirham in Gloucestershire, and which specimen he figures, though very wretchedly, for it appears to have been quite withered.

MAY 1ST, 1874.

This plant has been long cultivated at Kew, where it forms an erect shrub 3–4 feet high in the succulent house. The name *Anteuphorbium* was given because of its being a reputed antidote against the acrid poison of the Cape *Euphorbium*.

DESCR. An erect perfectly glabrous smooth pale green succulent shrub, with thick fleshy cylindric stem and branches, one half to one inch in diameter, which are constricted at the base. *Leaves* about an inch long, erect, oblong or linear-oblong, acute or obtuse, pale green and fleshy like the branches, with rounded quite entire margins; petiole excessively short, produced down the stem as three slender lines. *Heads* an inch long, cylindric, erect, solitary, axillary; peduncle very stout, almost clavate, with a few slender, scattered linear bracts, which are shorter than the head. *Involucral bracts* numerous, linear or linear-lanceolate, acuminate, green, slightly red at the base. *Flowers* all tubular, scarcely exceeding the involucre, yellow with a rose tinge. *Corolla* with short lobes. *Anthers* exserted. *Stigmatic arms* with short, conical, acute papillose tips. *Achene* small, quite smooth, crowned with a rather rigid pappus.—*J. D. H.*

Fig. 1, Vertical section of top of peduncle and head; 2, flower; 3, stamen: —*all magnified.*

TAB. 6100.

REGELIA CILIATA.

Native of South-Western Australia.

Nat. Ord. MYRTACEÆ.—Tribe LEPTOSPERMEÆ.

Genus REGELIA, *Schauer;* (*Benth. & Hook. f. Gen. Plant.*, vol. i. p. 706).

REGELIA *ciliata;* frutex hirsutus v. pubescens, foliis parvis erectis patentibus recurvisve late ovatis obovatis v. fere orbiculatis obtusis planis concavisve 3–5-nerviis, floribus in capitula globosa congestis, rachi lanata, calycis tubo ovoideo, lobis erectis, petalis minutis calycem vix excedentibus integerrimis ciliolatis, staminum phalangiis ungue lineari petala longe superantibus filamentis ad 12 erecto-patentibus filiformibus flabellatim dispositis, antheris minutis adnatis poris subterminalibus, calycibus fructiferis concretis ore lato truncato.

R. ciliata, *Schauer in Nov. Act. Nat. Cur.*, vol. xxi. p. 11, tab. i., figs. 1—3; *et in Plant. Preiss.*, vol. i. p. 148; *Benth. Flor. Austral.*, vol. iii. p. 170.

This genus, named after the distinguished and indefatigable Botanist and Superintendent of Culture in the Imperial Botanical Gardens of St. Petersburg, consists of three West Australian plants, which, with the habit of *Metrosideros*, are closely allied to *Beaufortia*, differing chiefly in the form of the anthers and number of ovules; by far the finest of them is the *R. grandiflora*, Benth., which has never yet been introduced into cultivation, and in which the apparently scarlet bundles of stamens are an inch long, and the leaves, which are many times larger than those of *R. ciliata*, and clothed with a white silky pubescence. All are greenhouse hard-wooded plants.

The species here figured has been cultivated for some years at Kew, flowering in September; and I have also received it in a flowering state from Messrs. Backhouse, of York. The ovary appears imperfect, as if the flower were male only. The fruiting specimen figured is from the Herbarium.

DESCR. A straggling twiggy bush, three to five feet high, with more or less pubescent or hirsute twigs and leaves; branchlets slender, strict, clothed with leaves. *Leaves* a quarter to a third of an inch long, erect, spreading, and

MAY 1ST, 1874.

recurved, imbricated tetrastichously, sessile, rigid, ovate obovate or almost orbicular, obtuse, flat or concave, quite entire, 3- rarely 5-nerved, hairy on both surfaces. *Flower-heads* globose, half to three-quarters of an inch in diameter, terminal, though as the axis elongates and becomes leafy the head becomes situated some inches below the top of the branchlet, of a dull red-purple colour. *Calyx-tube* villous, rounded at the base; lobes acute, subulate. *Petals* minute, scarcely exceeding the calyx, oblong, concave, quite entire, with ciliate margins. *Phalanges* of stamens much exceeding the calyx; claw linear, glabrous, giving off about twelve slender spreading filaments, each terminated by an adnate short two-celled anther, whose cells open by terminal short slits. *Ovary* a villous tubercle at the base of the calyx-tube.—*J. D. H.*

Fig. 1 and 2, Leaves; 3, flower; 4, longitudinal section of calyx; 5, petal; 6, tip of filament and anther:—*all magnified.*

TAB. 6101.

SENECIO DORONICUM, VAR. HOSMARIENSIS.

Native of Northern Marocco.

Nat. Ord. COMPOSITÆ.—Tribe SENECIONIDEÆ.

Genus SENECIO, *L.*; (*Benth. & Hook. f. Gen. Pl.*, vol. ii. p. 446).

SENECIO (Crociserides) *Doronicum;* herbaceus perennis polymorphus, floccoso-tomentosus v. glabratus, 1-pauci-cephalus, foliis crassiusculis dentatis, radicalibus lanceolatis ellipticis v. ovato-cordatis dentatis sinuato-dentatisve breviter v. longius petiolatis subacutis v. obtusis, involucri bracteati campanulati squamis lanceolatis acuminatis, ligulis 12-25 planis, achæniis glabris striatis.

S. Doronicum, *Linn. Sp. Pl.*, p. 1222; *DC. Prodr.*, vol. vi. p. 357.

VAR. *hosmariensis;* caule breviori, foliis caulinis paucis angustis, inferioribus 2-3 limbo lato in petiolum attenuatis, radicalibus late ovatis basi truncata v. subcordata in petiolum non decurrentibus. *Ball in Journ. Bot. n.s.*, vol. ii. p. 367 (1873).

Senecio Doronicum is a very handsome and not uncommon South European plant, extending from the Pyrenees to Transylvania, inhabiting considerable elevations in those countries, attaining a foot or two in height, with heads two inches in diameter. On the southern shores of the Mediterranean it has hitherto been found only in the northern mountains of Marocco, where it was discovered on Beni-Hosmar, a rugged limestone mass close to Tetuan, by Messrs. Ball, Maw, and myself, in April, 1871, at an elevation of about 3000 feet, growing in dry rocky places.

In this state it forms a very pretty rockwork plant, flowering in May in England. The specimen here figured is from Mr. Maw's rich garden of herbaceous plants at Benthall Hall, near Broseley, in Shropshire.

DESCR. A perennial scapigerous herb, leaves below, scape, bracts, and involucre more or less clothed with floccose tomentum. *Root* of thick fibres. *Radical-leaves* one to one and a half inches long, ovate, elliptic-ovate, or ovate-cordate, acute or obtuse irregularly toothed, more or less contracted

MAY 1ST, 1874.

into the short petiole or not, dark green, rugose, and glabrous above, greenish-white beneath. *Scape* three to five inches high, rather stout; bracts few, scattered, linear, foliaceous. *Heads* solitary, one and a half to two inches in diameter, yellow. *Involucre* campanulate, sub-biseriate; scales linear-subulate, green, with red purple tips, the innermost close-set, with tips slightly spreading, base woolly. *Ray-flowers* about twenty, tube glabrous, very broadly linear-oblong, tip three-toothed; disk-flowers shortly five-toothed, teeth obtuse erect. *Style-arms* truncate. *Achene* short, smooth, even, glabrous; pappus white, rather rigid, exceeding the involucre.—*J. D. H.*

Fig. 1, Ray-flower; 2, style-arms of ditto; 3, disk-flower; 4, style-arms:—*all magnified.*

Now Ready, Part III., with 4 Coloured Plates, Royal 4to, price 5s.

ORCHIDS,
AND

How to Grow them in India & other Tropical Climates.

BY

SAMUEL JENNINGS, F.L.S., F.R.H.S.,

Late Vice-President of the Agri-Horticultural Society of India.

NOW READY, Part II., 10s. 6d.

FLORA OF INDIA.

BY

DR. HOOKER, C.B., F.R.S.,

Assisted by various Botanists.

NOW READY, Vol. VI., 20s.

FLORA AUSTRALIENSIS.

A Description of the Plants of the Australian Territory. By GEORGE BENTHAM, F.R.S., assisted by BARON FERDINAND MUELLER, C.M.G., F.R.S. Vol. VI. Thymeleæ to Dioscorideæ.

NOW READY.

LAHORE TO YARKAND.

Incidents of the Route and Natural History of the Countries traversed by the Expedition of 1870, under T. D. FORSYTH, Esq., C.B. By GEORGE HENDERSON, M.D., F.L.S., F.R.G.S., Medical Officer of the Expedition, and ALLAN O. HUME, Esq., C.B., F.Z.S., Secretary to the Government of India. With 82 Coloured Plates of Birds and 6 of Plants, 26 Photographic Views of the Country, a Map of the Route, and Woodcuts. Price 42s.

In the Press and shortly to be published, in one large Volume, Royal 8vo, with numerous Coloured Plates of Natural History, Views, Map and Sections. Price 42s.

To Subscribers forwarding their Names to the Publishers before publication, 36s.

ST. HELENA:

A

Physical, Historical, and Topographical Description of the Island.

INCLUDING ITS

GEOLOGY, FAUNA, FLORA, AND METEOROLOGY.

BY

JOHN CHARLES MELLISS, C.E., F.G.S., F.L.S.

LATE COMMISSIONER OF CROWN PROPERTY, SURVEYOR AND ENGINEER OF THE COLONY.

L. REEVE & CO., 5, HENRIETTA STREET, COVENT GARDEN.

RE-ISSUE of the THIRD SERIES of the BOTANICAL MAGAZINE.

Now ready, Vols. I. to V., price 42s. each (to Subscribers for the entire Series 36s. each).

THE BOTANICAL MAGAZINE, Third Series. By Sir WILLIAM and DR. HOOKER. To be continued monthly.

Subscribers' names received by the Publishers, either for the Monthly Volume or for sets to be delivered complete at 36s. per volume, as soon as ready. The first complete set will be ready in a few days; others will speedily follow.

L. REEVE AND CO., 5, HENRIETTA STREET, COVENT GARDEN.

BOTANICAL PLATES;

OR,

PLANT PORTRAITS.

IN GREAT VARIETY, BEAUTIFULLY COLOURED, 6d. and 1s. EACH.

Lists of 2000 Species, one stamp.

L. REEVE AND CO., 5, HENRIETTA STREET, COVENT GARDEN.

SECOND-HAND.—THE FLORAL CABINET AND MAGAZINE OF EXOTIC BOTANY, by KNOWLES and WESTCOTT. Illustrated by 137 most beautifully Coloured Plates. Complete 3 vols. half morocco, price £2 2s. Published at £5 5s. unbound.

PLANTÆ ASIATICÆ RARIORES. By N. WALLICH, M. AND PH. D. Parts 2, 3, 4, and 5.

R. J. MITCHELL, 52, PARLIAMENT STREET.

QUADRANT HOUSE,

74, REGENT STREET, AND 7 & 9, AIR STREET, LONDON, W.

AUGUSTUS AHLBORN,

Begs to inform the Nobility and Gentry that he receives from Paris, twice a week, all the greatest novelties and specialties in Silks, Satins, Velvets, Shawls, &c., and Costumes for morning and evening wear. Also at his establishment can be seen a charming assortment of robes for Brides and Bridesmaids, which, when selected, can be made up in a few hours. Ladies will be highly gratified by inspecting the new fashions of Quadrant House.

From the *Court Journal*:—"Few dresses could compare with the one worn by the Marchioness of Bute at the State Concert at Buckingham Palace. It attracted universal attention, both by the beauty of its texture, and the exquisite taste with which it was designed. The dress consisted of a rich black silk tulle, on which were artistically embroidered groups of wild flowers, forming a most elegant toilette. The taste of the design, and the success with which it was carried out, are to be attributed to the originality and skill of Mr. AUGUSTUS AHLBORN."

Third Series.

No. 354.

VOL. XXX. JUNE. [*Price 3s. 6d. col^d. 2s. 6d. plain.*

OR No. 1048 OF THE ENTIRE WORK.

CURTIS'S
BOTANICAL MAGAZINE,

COMPRISING

THE PLANTS OF THE ROYAL GARDENS OF KEW,

AND OF OTHER BOTANICAL ESTABLISHMENTS IN GREAT BRITAIN,
WITH SUITABLE DESCRIPTIONS;

BY

JOSEPH DALTON HOOKER, M.D., C.B., F.R.S., L.S., &c.

Director of the Royal Botanic Gardens of Kew.

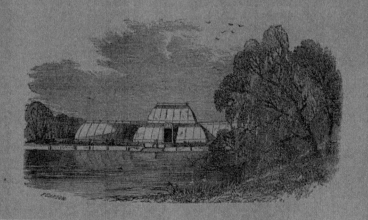

Nature and Art to adorn the page combine,
And flowers exotic grace our northern clime.

LONDON:
L. REEVE & CO., 5, HENRIETTA STREET, COVENT GARDEN.
1874.

[*All rights reserved.*]

ROYAL BOTANIC SOCIETY OF LONDON,
GARDENS—REGENT'S PARK.

ARRANGEMENTS FOR 1874.

SUMMER EXHIBITIONS, Wednesdays, June 10 and June 24. Gates open at 2 o'clock.

SPECIAL EVENING FETE, Wednesday, July 8. Gates open at 8 o'clock P.M. Evening Dress.

AMERICAN EXHIBITION, Daily, to June 9.

PROMENADES.—Every Wednesday in June and July, excepting the Exhibition days, commencing May 6. Visitors admitted only by the Special Coloured Orders.

LECTURES in the Museum at 4 o'clock precisely, Fridays, June 5, 12, 19, 26; July 3.

ROYAL HORTICULTURAL SOCIETY.

MEETINGS AND SHOWS IN 1874.

June 4 and 5.		(Great Summer Show.)
,,	17.	(Fruit and Floral Meeting.)
July	1.	(Cut Roses.)
,,	15.	(Zonal Pelargoniums.)
August	5.	(Fruit and Floral Meeting.)
,,	19.	Do.
September	2.	(Dahlias.)
October	7.	{ (Fruit and Floral Meeting.) { (Fungi.)
November	11.	(Fruit and Chrysanthemums.)
December	2.	(Fruit and Floral Meeting.)

RE-ISSUE of the THIRD SERIES of the BOTANICAL MAGAZINE.

Now ready, Vols. I. to VI., price 42s. each (to Subscribers for the entire Series 36s. each).

THE BOTANICAL MAGAZINE, Third Series. By Sir WILLIAM and DR. HOOKER. To be continued monthly.

Subscribers' names received by the Publishers, either for the Monthly Volume or for sets to be delivered complete at 36s. per volume, as soon as ready.

BOTANICAL PLATES;
OR,
PLANT PORTRAITS.

IN GREAT VARIETY, BEAUTIFULLY COLOURED, 6d. and 1s. EACH.

Lists of 2000 Species, one stamp.

L. REEVE & Co., 5, Henrietta Street, Covent Garden.

REDUCED IN PRICE.

SPECIES FILICUM; being Descriptions of the known Ferns, particularly of such as exist in the Author's Herbarium, accompanied with numerous Figures by Sir W. JACKSON HOOKER. 5 vols. 8vo, with 304 Plates. Bound in cloth extra, £7 8s., now offered for £4 4s.

DULAU AND Co., 37, Soho Square, London.

Tab. 6102.

SAXIFRAGA florulenta.

Native of the Maritime Alps.

Nat. Ord. Saxifragaceæ.—Tribe Saxifrageæ.

Genus Saxifraga, *L.*; (*Benth. & Hook. f. Gen. Pl.*, vol. i. p. 635).

Saxifraga (Euaizoonia) *florulenta;* foliis omnibus basalibus numerosissimis rosulatis confertis spathulatis coriaceis mucronatis glabris margine cartilagineo setaceo-ciliato apicem versus integerrimo, caule florifero erecto thyrsoideo fistuloso hirsuto, bracteis linearibus, pedunculis unifloris rarius bifloris erectis apice cernuis inferioribus longioribus, bracteolis sub flores fere 2 lanceolato-linearibus, calycis glanduloso-hispidi tubo obconico ovario adnato, lobis lanceolatis mucronulatis erectis, petalis atque staminibus duplo longioribus spathulatis obtusis 5-nerviis, stylis tribus capitatis staminibus æquilongis.

Saxifraga florulenta, *Moretti Tent. Sax.*, p. 9; *Seringe in DC. Prodr.*, vol. iv. p. 20; *Bertolon. Misc. Bot.*, xxi. p. 14, t. 2; *Engler Monog. der Gatt. Saxifr.*, p. 248; *Regel Gartenfl.* 1874, p. 2, t. 782.

This striking and extremely local species was first discovered about the year 1820 (in the Alps of Fenestre) by an English tourist, who forwarded specimens to Professor Moretti of Pavia. It was rediscovered in the same locality in the year 1856, since which time it has been found in numerous distinct habitats. It appears to be tolerably abundant at an altitude of from 7000 to 9000 feet within a limited area of about eight miles square, in the higher regions of the watershed of the Maritime Alps, between the Col du Tenda and the valley of the Tinea north of Nice, on cliff faces and precipitous ravines facing the north. Mr. G. Maw, to whom we are indebted for the specimen figured, informs me that it mostly grows in single rosettes, some of which are six or seven inches across; they are generally found under an overhanging ledge protected from the drip and direct rainfall, the rosette turning downwards, and never exposed to the sun. The plant was first introduced alive to this country by Mr. Moggridge. Its cultivation is extremely difficult, from the all but impossibility of obtaining well-rooted plants. It is an

June 1st, 1874.

extremely shy bloomer; it probably lives to a great age before flowering, after which it dies. It seems entirely to fail under pot culture, but Mr. Maw informs me that M. Boissier has succeeded in growing it by wedging the rosettes firmly into the crevices of a brick wall with a northern exposure. Mr. Ellacombe has found it intolerant of frost at Bitton, near Bristol.

DESCR. *Rosettes* five to seven inches in diameter, concave, becoming convex at the time of flowering, bright green. *Leaves* three-quarters of an inch to two inches long, innermost shortest, very numerous, densely imbricated, spathulate, mucronate, margin cartilaginous, with setaceous cilia below, entire towards the apex. *Inflorescence* a narrow thyrsoid panicle with a fistulose rachis, five to twelve inches high, more or less densely hairy; bracts linear-spathulate; peduncles 1–2-flowered, with one or two linear-lanceolate bracteoles. *Flowers* half an inch long, slightly nodding. *Calyx* obconical, densely clothed with gland-tipped hairs; lobes lanceolate, mucronulate. *Petals* pale lilac, twice as long as the calyx-lobes and stamens, spathulate, obtuse, 5-nerved. *Ovary* three-celled; styles capitate. *Capsule* globular.—*W. T. T. D.*

Fig. 1, Leaf of a rosette; 2, do. back; 3, transverse section; 4, flower; 5, do. calyx, limb, and petals removed; 6, transverse section of ovary:—*all magnified.*

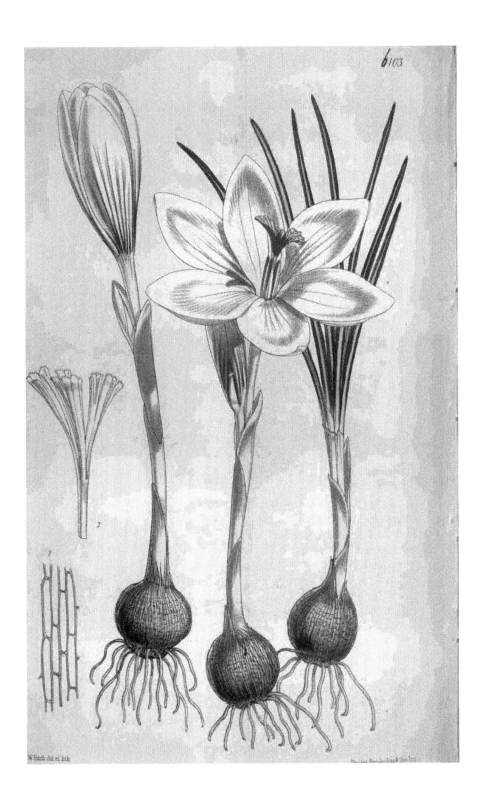

Tab. 6103.

CROCUS CANCELLATUS.

Native of Greece and Asia Minor.

Nat. Ord. IRIDEÆ.—Tribe IXIEÆ.

Genus CROCUS, *Tourn.; (Klatt in Linnæa,* vol. xxxiv. p. 674).

CROCUS *cancellatus;* cormis mediocriter globosis, tunicis reticulatis fibrosis brunneis demum in fibras setiformes perianthii tubi basim circumdantes solutis, areolis elliptico-oblongis, foliis post anthesin evolutis 6-8-pollicaribus, vaginis latis membranaceis, spatha bivalvi vel monophylla, involucro, perianthii tubo albido vel pallide lilacino, segmentis albis, vel lilacinis purpureo striatis ellipticis acutiusculis, fauce lutea levi, antheris aureis filamentis levibus longioribus, stigmatibus croceis multifidis antheras superantibus.

CROCUS cancellatus, *Herbert in Bot. Mag.* sub t. 3864; *Journ. Hort. Soc.,* vol. ii. p. 276; *Baker in Gard. Chron.,* 1873, p. 1553.

C. Schimperi, *Gay in Schimp. Pl. Cephal. exsic.*

C. Spruneri, *Boiss. & Heldr. Diagn.,* vii. p. 103.

C. mazziaricus, *Herb. Bot. Reg.* 1845, *Misc.,* p. 3; *Bot. Reg.,* 1847, t. 16, f. 5 & 6.

C. dianthus, *K. Koch in Linnæa,* vol. xxi. p. 634.

C. nudiflorus, *Sibth. & Smith, Prodr. Fl. Græc.,* p. 23 *(excl. syn.).*

A well-known and beautiful species not apparently figured, except very ineffectively by Herbert. It is frequent in the Ionian Islands, Greece, Asia Minor, and reaches, it is said, eastward to Armenia. In Greece it ascends to 4000 feet and in Taurus to the Alpine region. The curious reticulated coats, with a texture which recalls cocoa-nut fibre, especially in the prolonged bristle-like fibres surrounding the neck (similar to those met with in *Albuca setosa* or *Chlorogalum pomeridianum*) distinguish this species from all the autumn flowering Crocuses. Amongst the spring-flowering species, coats of this kind are only found in *C. reticulatus* and *C. susianus.* The specimens figured were received from G. Wood, Esq., of Rochford, Essex, early in November, 1873.

DESCR. *Corm* globose, about one inch in diameter, clothed with coarse reticulated fibrous coats (not well represented in

JUNE 1ST, 1874.

the plate) prolonged above into loose bristle-like fibres. *Leaves* about seven, produced after the flowers (half their length in the plate). *Perianth* without an involucre; tube slender, pale; throat pale yellow; limb three and a half inches in diameter, segments elliptic rather acute white with reddish-purple streaks. *Anthers* shorter than the filaments. *Stigma* saffron coloured, divided into numerous segments.—*W. T. T. D.*

Fig. 1, Reticulated coat of the corm; 2, stigmas:—*both magnified.*

Tab. 6104.

CALANTHE curculigoides.

Native of Malacca.

Nat. Ord. Orchideæ.—Tribe Vandeæ.

Genus Calanthe, *Br.*; (*Lindl. Fol. Orchid. Calanthe*).

Calanthe (Eucalanthe) *curculigoides;* foliis anguste ellipticis glabris scapo duplo longioribus, racemo cylindraceo denso, bracteis membranaceis caducis, labello trilobo, lobis lateralibus rotundatis, medio subpanduriformi acuto nervis quinque parallelis, calcare uncinato, rostello longo setiformi faucem occludente.

Calanthe curculigoides, *Lindl. Gen. & Sp. Orch.*, p. 251; *Bot. Reg.* 1847, t. 8; *Fol. Orchid. Calanthe*, p. 4; *Walp. Ann.*, vol. i. p. 792 *et* vol. vi. p. 913; *Wall. Cat.*, 7340.

Styloglossum nervosum, *Kuhl & Hasselt* ex *Herb. Lindl.*; *Breda Orch. Jav.*, t. 7.

? Amblyglottis pulchra, *Blume Bijd.* 371.

? Calanthe pulchra, *Lindl. Gen. & Sp. Orch.*, p. 250.

A native of Malacca, Penang, and Singapore, according to Lindley, who described it originally from specimens collected in the two latter countries by Wallich, and who afterwards figured it in the *Botanical Register* from a plant flowered by the Messrs. Loddiges of Hackney. In the latter work Dr. Lindley describes it as an undisputed novelty by reason of its colour, which is pale dirty ochreous yellow in his figure, with much larger flowers than our plant, in which they are of a rather bright orange. The flowers too are much more ringent in Lindley's figure and dried specimen, and the lip is more acute, but I can hardly think the species are different. I have taken the two doubtful synonyms from the *Botanical Register*, having no means of verifying them. Lindley also gives Java as a habitat.

Descr. Terrestrial. *Leaves* subradical, distichous, sheathing at the base, narrowly elliptic or elliptic-lanceolate, acute, with about six strong parallel nerves, deep green above, paler beneath. *Racemes* three to four inches long (or more), cylin-

June 1st, 1874.

drical, terminating a lateral scape about half the length of the leaves, and furnished with several brown sheathing scales. *Bracts* linear, membranous, brown, very fugitive. *Flowers* orange yellow, three quarters of an inch long, crowded; pedicel one to three quarters of an inch. *Perianth* scarcely patent. *Sepals* oblanceolate, acute. *Petals* similar. *Lip* 3-lobed, about equalling the perianth; lateral lobes rounded, intermediate lobe subpanduriform, acute. *Rostellum* a slender style-like process half as long as the lip, and partially closing the mouth of the hooked spur.—*J. D. H.*

Fig. 1, Flower; 2, the same with the sepals and petals removed; 3, lip, column, and base of ovary; 4, pollinia:—*all magnified.*

Tab. 6105.

GREVILLEA FASCICULATA.

Native of West Australia.

Nat. Ord. PROTEACEÆ.—Tribe GREVILLEÆ.

Genus GREVILLEA, *R. Br.*; (*Benth. Fl. Austr.*, vol. v. p. 417).

GREVILLEA (Plagiopoda) *fasciculata;* frutex, ramulis pubescentibus, foliis pollicaribus sesquipollicaribusve sessilibus vel brevissime petiolatis oblongo- vel oblanceolato-linearibus acutis margine revolutis supra scabro-punctatis, racemis terminalibus axillaribusve sessilibus paucifloris, perianthii pilosiusculi tubo basi gibboso sub limbo globoso reflexo, fauce pilis reflexis transverse barbata, disci valde obliqui glandula latissima truncata, ovario stipitato piloso, stigmate obliquo crasso concavo glabro.

GREVILLEA fasciculata, *R. Br. Suppl.*, p. 20; *Meissn. in Pl. Preiss.*, vol. i. p. 536; *DC. Prodr.* vol. xiv. p. 369; *Benth. Fl. Austral.*, vol. v. p. 449.

G. Meissneriana, *F. Muell. in Linnæa*, vol. xxvi. p. 357; *Meissn. l.c.* p. 360.

G. aspera *var.* linearis, *Meissn. in Pl. Preiss.*, vol. i. p. 537.

A native of Western Australia, from King George's Sound to the Swan River. It was discovered on the south-west shore of the former place in 1829, by D. Baxter. The specimen figured was from a Swan River plant, flowered by Mr. Wilson Saunders, at Reigate, in May, 1873.

DESCR. A low and prostrate or erect shrub, attaining three to four feet. *Branches* slender, erect, more or less fasciculate, leafy, terete, youngest pubescent. *Leaves* three-quarters to one and a-half inches long, sessile, or very nearly so, acute, somewhat variable in form from oblanceolate to linear-oblanceolate and linear, rather rigid with revolute margins, upper surface minutely scabrous-punctate, lower silky and pale brown. *Racemes* or *fascicles* few-flowered, sessile, axillary and terminal on the lateral shoots. *Flowers* bright red with yellow tips, pedicels about half their length. *Perianth* a quarter to a third of an inch long, sac-like and gibbous at the base, contracted and revolute below the yellow limb, the tube furnished internally on the superior side with a transverse beard of

JUNE 1ST, 1874.

reflexed hairs below the contraction. *Torus* oblique, gland broad, truncate. *Ovary* very shortly stipitate, hairy. *Style* dilated upwards, somewhat exceeding the perianth, stigma very oblique glabrous fleshy concave and slightly hippocrepiform.—*W. T. T. D.*

Fig. 1, Leaf, under surface; 2, ditto, upper surface; 3, flower, lateral view; 4, ditto, longitudinal section:—*all magnified.*

Tab. 6106.

LESSERTIA PERENNANS.

Native of Natal.

Nat. Ord. LEGUMINOSEÆ.—Tribe GALEGEÆ.

Genus LESSERTIA, *DC.*; (*Benth. & Hook. f. Gen. Pl.*, vol. i. p. 503).

LESSERTIA *perennans;* suffrutex, caulibus erectis virgatis valde striatis tenuiter puberulis, foliolis 7-10-jugis ellipticis elliptico-lanceolatisve utrinsecus sericeo-pubescentibus, racemis laxiusculis folia superantibus, floribus lilacinis, calycis lobis attenuatis, legumine stipitato compresso oblique elliptico mucronato glabro, seminibus 3-4.

LESSERTIA perennans, *DC. Prodr.,* vol. ii. p. 271; *Harv. & Sond. Fl. Cap.,* vol. ii. p. 216.

All the species of the genus are natives of Southern Africa. The present one occurs in grassy places on the eastern side, from Albany to the Transvaal. The specimen figured was from a plant which flowered in the Royal Gardens, Kew, in July, 1873.

DESCR. Suffruticose with erect virgate strongly striate slightly pubescent fistular stems. *Leaves* imparipinnate, about two inches long, more or less silkily pubescent on both surfaces, and greyish, or rarely glabrate; leaflets seven to ten pairs, very shortly petioluled, elliptic or elliptic-lanceolate, acute or mucronulate; petiole about a quarter of an inch long; stipules linear-falcate, membranous. *Racemes* rather lax, exceeding the leaves; peduncle one to three inches long. *Pedicels* rather longer than the flowers. *Flowers* a quarter of an inch long, reddish-lilac or white. *Calyx* pubescent; teeth narrowly triangular, gradually attenuate. *Legumes* about three quarters of an inch long, obliquely elliptic or oblong-elliptic, very shortly stipitate, compressed, membranous, 3-4-seeded—*W. T. T. D.*

Fig. 1, Leaflet; 2, flower; 3, do. petals removed; 4, standard; 5, wing; 6, keel; 7, ovary:—*all magnified.*

Now Ready, Part IV., with 4 Coloured Plates, Royal 4to, price 5s.

ORCHIDS,

AND

How to Grow them in India & other Tropical Climates.

BY

SAMUEL JENNINGS, F.L.S., F.R.H.S.

Late Vice-President of the Agri-Horticultural Society of India.

NOW READY, Part II., 10s. 6d.

FLORA OF INDIA.

BY

DR. HOOKER, C.B., F.R.S.,

Assisted by various Botanists.

NOW READY, Vol. VI., 20s.

FLORA AUSTRALIENSIS.

A Description of the Plants of the Australian Territory. By GEORGE BENTHAM, F.R.S., assisted by BARON FERDINAND MUELLER, C.M.G., F.R.S. Vol. VI. Thymeleæ to Dioscorideæ.

NOW READY,

LAHORE TO YARKAND.

Incidents of the Route and Natural History of the Countries traversed by the Expedition of 1870, under T. D. FORSYTH, Esq., C.B. By GEORGE HENDERSON, M.D., F.L.S., F.R.G.S., Medical Officer of the Expedition, and ALLAN O. HUME, Esq., C.B., F.Z.S., Secretary to the Government of India. With 32 Coloured Plates of Birds and 6 of Plants, 26 Photographic Views of the Country, a Map of the Route, and Woodcuts. Price 42s.

In the Press and shortly to be published, in one large Volume, Royal 8vo, with numerous Coloured Plates of Natural History, Views, Map and Sections. Price 42s.

To Subscribers forwarding their Names to the Publishers before publication, 36s.

ST. HELENA:

A

Physical, Historical, and Topographical Description of the Island,

INCLUDING ITS

GEOLOGY, FAUNA, FLORA, AND METEOROLOGY.

BY

JOHN CHARLES MELLISS, C.E., F.G.S., F.L.S.

LATE COMMISSIONER OF CROWN PROPERTY, SURVEYOR AND ENGINEER OF THE COLONY.

L. REEVE AND Co., 5, Henrietta Street, Covent Garden.

DEDICATED BY SPECIAL PERMISSION TO H.R.H. THE
PRINCESS OF WALES.

NOW READY, Complete in Six Parts, 21s. each, or in One Vol., imperial, folio, with 30 elaborately Coloured Plates, forming one of the most beautiful Floral Works ever published, half morocco, gilt edges, £7 7s.

A MONOGRAPH OF ODONTOGLOSSUM.

A Genus of the Vandeous section of Orchidaceous Plants. By JAMES BATEMAN, F.R.S., F.L.S., Author of "The Orchidaceæ of Mexico and Guatemala."

L. REEVE & Co., 5, Henrietta Street, Covent Garden.

NOW READY, Vol. 3, with 551 Wood Engravings, 25s.

THE NATURAL HISTORY OF PLANTS.

By Prof. H. BAILLON, P.L.S., Paris. Translated by MARCUS M. HARTOG, B.Sc., Lond., B.A., Scholar of Trinity College, Cambridge. Contents :—Menispermaceæ, Berberidaceæ, Nymphæaceæ, Papaveraceæ, Capparidaceæ, Cruciferæ, Resedaceæ, Crassulaceæ, Saxifragaceæ, Piperaceæ, Urticaceæ.

L. REEVE & Co., 5, Henrietta Street, Covent Garden.

QUADRANT HOUSE,
74, REGENT STREET, AND 7 & 9, AIR STREET, LONDON, W.

AUGUSTUS AHLBORN,

Begs to inform the Nobility and Gentry that he receives from Paris, twice a week, all the greatest novelties and specialties in Silks, Satins, Velvets, Shawls, &c., and Costumes for morning and evening wear. Also at his establishment can be seen a charming assortment of robes for Brides and Bridesmaids, which, when selected, can be made up in a few hours. Ladies will be highly gratified by inspecting the new fashions of Quadrant House.

From the *Court Journal*:—"Few dresses could compare with the one worn by the Marchioness of Bute at the State Concert at Buckingham Palace. It attracted universal attention, both by the beauty of its texture, and the exquisite taste with which it was designed. The dress consisted of a rich black silk tulle, on which were artistically embroidered groups of wild flowers, forming a most elegant toilette. The taste of the design, and the success with which it was carried out, are to be attributed to the originality and skill of Mr. AUGUSTUS AHLBORN.

Third Series.

No. 355.

VOL. XXX. JULY. [*Price 3s. 6d. col*^{d.} *2s. 6d. plain.*

OR No. 1049 OF THE ENTIRE WORK.

CURTIS'S
BOTANICAL MAGAZINE,

COMPRISING

THE PLANTS OF THE ROYAL GARDENS OF KEW,

AND OF OTHER BOTANICAL ESTABLISHMENTS IN GREAT BRITAIN,
WITH SUITABLE DESCRIPTIONS;

BY

JOSEPH DALTON HOOKER, M.D., C.B., F.R.S., L.S., &c.

Director of the Royal Botanic Gardens of Kew.

Nature and Art to adorn the page combine,
And flowers exotic grace our northern clime.

LONDON:
L. REEVE & CO., 5, HENRIETTA STREET, COVENT GARDEN.
1874.

[*All rights reserved.*]

ROYAL BOTANIC SOCIETY OF LONDON,

GARDENS—REGENT'S PARK.

ARRANGEMENTS FOR 1874.

SPECIAL EVENING FETE, Wednesday, July 8. Gates open at 8 o'clock P.M. Evening Dress.

PROMENADES.—Every Wednesday in July. Visitors admitted only by the Special Coloured Orders.

LECTURE in the Museum at 4 o'clock precisely, Friday, July 3.

ROYAL HORTICULTURAL SOCIETY.

MEETINGS AND SHOWS IN 1874.

July	1.	(Cut Roses.)
,,	15.	(Zonal Pelargoniums.)
August	5.	(Fruit and Floral Meeting.)
,,	19.	Do.
September	2.	(Dahlias.)
October	7.	{ (Fruit and Floral Meeting.) (Fungi.)
November	11.	(Fruit and Chrysanthemums.)
December	2.	(Fruit and Floral Meeting.)

RE-ISSUE of the THIRD SERIES of the BOTANICAL MAGAZINE.

Now ready, Vols. I. to VII., price 42s. each (to Subscribers for the entire Series 36s. each).

THE BOTANICAL MAGAZINE, Third Series. By Sir WILLIAM and DR. HOOKER. To be continued monthly.

Subscribers' names received by the Publishers, either for the Monthly Volume or for sets to be delivered complete at 36s. per volume, as soon as ready.

BOTANICAL PLATES;
OR,
PLANT PORTRAITS.

IN GREAT VARIETY, BEAUTIFULLY COLOURED, 6d. and 1s. EACH.

Lists of 2000 Species, one stamp.

L. REEVE & Co., 5, Henrietta Street, Covent Garden.

THE FLORAL MAGAZINE.

NEW SERIES, ENLARGED TO ROYAL QUARTO.

Figures and Descriptions of the Choicest New Flowers for the Garden, Stove, or Conservatory. Monthly, with 4 Coloured Plates, 3s. 6d. Annual Subscription, 42s.

TAB. 6107.

CHRYSANTHEMUM CATANANCHE.

Native of the Greater Atlas.

Nat. Ord. COMPOSITÆ.—Tribe ANTHEMIDEÆ.

Genus CHRYSANTHEMUM, *L.*; (*Benth. & Hook. f. Gen. Plant.*, vol. ii. p. 424).

CHRYSANTHEMUM *Catananche;* perenne, subcæspitosum, sericeo-pilosum v. tomentosum, foliis radicalibus fasciculatis inæqualiter 1-3-ternatim sectis segmentis patentibus linearibus acutis obtusisve, petiolo lineari, scapis adscendentibus superne nudis 1-cephalis, capitulo 1–2 poll. diam., involucri bracteis laxe imbricatis oblongis medio herbaceis late scarioso-marginatis nitidis, ligulis ad 25 latiusculis stramineis basi sanguineis obtuse 3-dentatis, disci floribus flavis, achæniis lineari-oblongis 10-costatis costis anguste alatis.

CHRYSANTHEMUM Catananche, *Ball in Trimen. Journ. Bot.*, 1873, p. 366.

This, which is one of the most beautiful plants of the Greater Atlas, was discovered in 1871, by Messrs. Ball, Maw, and myself, in valleys of that range at elevations of 7000 to 9000 feet, flowering in May, and has since been cultivated both in Mr. Maw's garden, and at Kew, where it flowered for the first time in April of the present year. In its native country it forms patches of a silvery green hue, and of considerable size, in rocky valleys, and on mountain slopes exposed to the sun. The broad white involucral bracts are conspicuous for their silvery whiteness, hyaline texture, and transparency, relieved by a narrow purplish herbaceous central band; their resemblance to the bracts of *Catananche* has suggested the specific name.

DESCR. *Rootstock* stout, woody, branched, with often many heads. *Leaves* tufted, one to two and a half inches long, as well as the scape clothed with silky tomentum of a silvery green colour, petioled, irregularly 3-chotomously cut once twice or thrice into linear acute or obtuse spreading lobes; petiole slender, narrow. *Scapes* stout, ascending, three to six inches high, green. *Heads* solitary, one and a half to two inches across, pale yellow, the rays of dirty

JULY 1ST, 1874.

purplish hue outside towards the tip, and blood-red within at the very base; disk of a darker yellow. *Involucre* campanulate; bracts imbricate, linear-oblong, scarious, white and transparent, with a purplish herbaceous median band. *Ray-flowers* about twenty-five, with a broad linear-oblong obtusely 3-toothed ligula and short glabrous tube; style-arms oblong, obtuse; achene linear, deeply 10-ribbed, ribs with narrow membranous wings; pappus short, membranous, obliquely truncate. *Disk-flowers* slender, 5-toothed, glabrous, style-arms with broad truncate tips; achenes narrow, like those of the ray but shorter; pappus a short auricle—*J. D. H.*

Fig. 1, Ray; and 2, disk-flowers:—*both magnified*.

Tab. 6108.

ERICA Chamissonis.

Native of South Africa.

Nat. Ord. ERICACEÆ.—Tribe ERICEÆ.

Genus ERICA, *L.*; (*Benth. in DC. Prodr.*, vol. vii. p. 613).

ERICA (Melastemon) *Chamissonis*; erecta, ramosa, hirto-pubescens, foliis incurvi-patentibus 3-nis ⅓ poll. longis anguste linearibus dorso sulcatis, floribus numerosis in ramulis abbreviatis terminalibus, pedicellis hirtis, bracteis minutis, calycis parvi segmentis acuminatis, corolla globoso-campanulata, lobis latis brevibus, staminibus inclusis filamentis brevibus glabris, antheris brevibus, loculis apice subacutis lateraliter anguste cristatis poris amplis apices versus lateralibus, ovario hirto, stylo gracili, stigmato truncato.

ERICA Chamissonis, *Klotzsch in Herb. Reg. Berol. ex Benth. in DC. Prodr.*, vol. vii. p. 685.

Many years ago the Cape Heaths formed a conspicuous feature in the greenhouses of our grandfathers, and in the illustrated horticultural works of the day, including this Magazine, wherein about 50 are figured. These have given place to the culture of soft-wooded plants—Geraniums, Calceolarias, Fuchsias, &c.; and the best collections of the present day are mere ghosts of the once glorious Ericeta of Woburn, Edinburgh, Glasgow, and Kew. A vast number of the species have indeed fallen out of cultivation, and a few easily propagated hybrids for decorative purposes are all that are to be seen of this lovely tribe in most of the best establishments of England. No less than 186 species of *Erica* were cultivated at Kew in the year 1811, now we have not above 50, together with many hybrids and varieties. Besides the fact of their going out of fashion, there have been two main causes for their present rarity; of these the first and most conspicuous is bad treatment. As with Australian and other Cape hard-wooded plants, their culture is special, unknown to most gardeners of the present day, and they will not survive the promiscuous use of the water-pot and syringe, to which they are exposed if mixed up with many other things.

The second is, that very few collectors have been of late years in the Heath district of the Cape, which is almost confined to the narrow strip of country between the Western coast and the coast ranges, and where were the botanizing grounds of the collectors sent out at the beginning of the century.

Erica Chamissonis is one of the few Heaths that extend eastward in South Africa, being found near Graham's Town in the Albany district, about 500 miles east of Cape Town, where it grows on rocky hills at an elevation of 2000 feet, flowering in October. Seeds of it were sent to the Royal Gardens by Mr. M'Owan. The plant here figured, raised from these, flowered in April.

DESCR. A shrub with slender leafy erect branches, all parts, except the corolla, clothed with short soft spreading hairs. *Leaves* about a quarter to a third of an inch long, ternate, spreading and incurved, sessile, linear, obtuse, grooved underneath from the recurvation of the margin. *Flowers* at the tips of short side-branches, solitary or three or four together, pendulous, rose-coloured, about a third of an inch in diameter; pedicel half an inch long, pink, with two small basal bracts and two bracteoles above them. *Calyx* jointed with the pedicel, small; teeth ovate, acuminate, much shorter than the corolla. *Corolla* between globose and campanulate; lobes very short and broad. *Stamens* short, filaments glabrous; anthers short, with narrowly crested pointed cells and lateral slits near the tip. *Ovary* tomentose, 4-celled; style slender, stigma truncate; ovules many in each cell. —*J. D. H.*

Fig. 1, Leaves; 2, flowers; 3, the same with the corolla removed; 4 and 5, stamens; 6, ovary; 7, transverse section of do. :—*all magnified.*

TAB. 6109.

ROMANZOFFIA SITCHENSIS.

Native of North-West America.

Nat. Ord. HYDROLEACEÆ.—Tribe NAMEÆ.

Genus ROMANZOFFIA, *Cham.;* (*Choisy in DC. Prodr.*, vol. x. p. 185).

ROMANZOFFIA *sitchensis;* tota pilis crispulis aspersa, foliis reniformi-cordatis suborbiculatisve crenato-lobatis, cymis laxifloris.

ROMANZOFFIA sitchensis, *Chamiss. in Linnœa*, vol. ii. p. 609; *Bongard Bot. Sitch.*, p. 41, t. 4; *Hook. Fl. Bor. Am.*, vol. ii. p. 103; *Ledeb. Flor. Ross.*, vol. iii. p. 181; *Regel Gartenfl.*, vol. xxii. (1873) p. 33, t. 748.

This very rare and interesting little plant, with the habit of a Saxifrage of the *granulata* group, is closely allied to the majestic *Wigandia* of our subtropical gardens, though so dissimilar in stature, habit, and general characters, and in coming from so different a climate and country. It is a native of a few distant spots over a very wide range of country in North-Western America, and has been gathered by very few collectors. First, by the late venerable Menzies, the Naturalist to Vancouver's voyage (and introducer of *Araucaia imbricata*) in May, 1793, who discovered a small slender variety of it on hanging rocks at Trinidad, in California, lat. 41° 10′ N.; next by Chamisso at Sitka in the then Russian, but now American territory of Alaschka, fully 1000 miles north of Trinidad, and by whom it was first described; more lately it was gathered abundantly by Dr. Lyall on the Cascade Mountains, in lat. 69° N. in the bed of the Sallse river, and a large flowered variety (Regel's *R. grandiflora*) on the same mountains, at an elevation of 7000 feet. Lastly we have specimens collected in South California (probably in the mountains), in lat. 35°, by Dr. Bigelow, surgeon to Lieutenant Whipple's exploration for a railway route across America in 1853-4; this is fully 1400 miles south of Sitka.

Romanzoffia sitchensis is a rock-plant, easy of cultivation,

JULY 1ST, 1874.

and was, I believe, introduced into Europe by Messrs. Haage and Schmidt, of Erfurt. The specimen here figured flowered in the Royal Gardens in April last.

DESCR. A weak, green, perennial-rooted, straggling, or sub-erect herb, four to eight inches high or long, more or less covered with scattered curled short hairs. *Stems* many from the root, branched. *Leaves* subradical, petioled, one to one and a half inches in diameter, orbicular-reniform, crenate-lobed, bright green, paler beneath; petiole a half to one inch long. *Cymes* at the ends of the branches, few-flowered, ebracteate. *Flowers* variable in size, one-third to one-half inch in diameter, white; pedicels slender, spreading. *Sepals* oblong-ovate, sub-acute. *Corolla-lobes* orbicular. *Stamens* attached to the base of the corolla-tube. *Disk* annular. *Ovary* glabrous, style slender, stigma minute.—*J. D. H.*

Fig. 1, Flower; 2, corolla laid open; 3, disk and ovary; 4, transverse section of ovary :—*all magnified.*

TAB. 6110.

IRIS OLBIENSIS.

Native of Northern Italy and Southern France.

Nat. Ord. IRIDACEÆ.—Tribe IRIDEÆ.

Genus IRIS, *Linn.*; (*Endl. Gen. Plant.*, p. 266).

IRIS *olbiensis;* rhizomate crasso, caule brevi, foliis brevibus (2-6 poll.) latiusculis acutis sensim acuminatis scapo brevioribus, floribus breviter pedicellatis magnis, spathæ valvis membranaceis laxis abrupte acuminatis ovarium velantibus, perianthii tubo pollicari, limbi 3 poll. lati segmentis obovato-spathulatis decurvis apice rotundatis ungue barbato, interioribus iis subæqualibus erecto-incurvis conniventibus elliptico-oblongis stipitatis, stigmatibus segmentis perianthii dimidio brevioribus bifidis lobis triangularibus acutis margine exteriore dentatis.

IRIS olbiensis, *Hénon in Ann. Soc. Agric. Lyons; Gren. et Godr. Fl. de France,* vol. iii. pt. i. p. 240; *Parlatore Flor. Ital.,* vol. iii. p. 283.

This belongs to a small group of dwarf Iris, which inhabit for the most part Southern Europe, and of which the *I. pumila,* L. (Tab. nost. 9, 1209 and 1261) may be taken as the type. It is a native of the South of France and North Italy, from Nismes eastwards, but apparently not advancing beyond Tuscany. It varies much in the colour of the flowers, which are sometimes white. It is distinguished from *I. pumila* by the much larger flowers, which are pedicelled and less fugacious, as also by the shorter perianth-tube. The *I. italica* of Parlatore appears to be only a variety of it; and it is represented by *I. pseudo-pumila* in Sicily. It is very closely allied to, if not a mere variety of the *I. Chamaciris,* Bertoloni, which has a wider range in France and Italy. The specimen here figured flowered in the Royal Gardens in April of the present year.

DESCR. *Rootstock* prostrate, very thick and fleshy, as big as the thumb. *Leaves* three to six inches long by one-third to two-thirds inch in diameter, erecto-patent, straight or somewhat falcate, usually narrowed almost from the base to the acuminate tip, glaucous green. *Scape* rather larger than the

JULY 1ST, 1874.

leaves, stout, erect, closely sheathed. *Spathes* one to two inches long, large and rather tumid, lax, obliquely truncate and acuminate. *Flowers* very large for the size of the plant, usually dark purple, three and a half to four inches across the perianth were it spread out; pedicel short, stout. *Perianth-tube* longer than the ovary; outer segments recurved, spathulate-obovate, tip rounded, claw deeply bearded; inner segments as long as the outer, erect and connivent, broadly elliptic-oblong with a narrow claw and cordate base. *Stigmas* not half the length of the inner perianth segments, their lobes triangular, acute, toothed at the outer edge.—*J. D. H.*

TAB. 6111.

CAMPSIDIUM CHILENSE.

Native of Chili.

Nat. Ord. BIGNONIACEÆ.—Tribe BIGNONIEÆ.

Genus CAMPSIDIUM, *(Seemann in Bonplandia,* vol. x. (1862), p. 147).

CAMPSIDIUM *chilense;* frutex volubilis glaberrimus, foliis oppositis impari-pinnatis, foliolis oppositis ellipticis v. ovato- v. elliptico- v. lanceolato-oblongis obtusis v. apiculatis integerrimis serratisve rachi antice sulcato, racemis terminalibus pendulis 6-10-floris, floribus coccineis gracile pedicellatis.

CAMPSIDIUM chilense; *Reiss & Seem.; ex Seem. in Bonplandia,* vol. x. p. 147, t. 11; *Gard. Chron.,* 1870, p. 1182, *cum ic xylog.*

TECOMA Guarume, *Hook. in Bot. Mag.,* t. 4896 *in adnot.* (*non DC.*)

T. valdiviana, *Philippi in Linnæa,* 1857, p. 14.

T. mirabilis, *Hort.*

This very beautiful climber is a native of Chili and the Archipelago of Chiloe, and was discovered on the island of Huaffo by Dr. Eights, an American voyager, who sent a small collection of Chilian and Fuegian plants to Sir William Hooker some fifty years ago, amongst which is this plant. It has subsequently been collected by many botanists, most recently by Dr. Cunningham, naturalist to the surveying expedition of H.M.S. *Nassau,* who gathered it as far south as Wellington Island in lat. 40° S., where it would seem to be common. Its northern limit is probably Arique, near Valdivia, lat. 50° S., where it was found by Lechler. It is not a little remarkable that so beautiful a plant, and one found through so many degrees of latitude in Chili, should have escaped the observation of C. Gay, whose Flora Chilensis, published in 1845, does not include it. The equally conspicuous *Berberidopsis corallina* (Tab. nost. 5343) which, like *Campsidium,* is a native of the neighbourhood of the maritime capital of Valdivia, was also unknown to that author, though he spent many years exploring that country for the Chilian government. I am indebted to Messrs. Veitch for the plant here

JULY 1ST, 1874.

figured, which flowered with them in April of the present year.

DESCR. A woody perfectly glabrous slender climber, ascending trees to a height of forty to fifty feet; branches woody, angular, with pale yellowish bark, wood very hard. *Leaves* four to six inches long; leaflets three-quarters to one and a half inches long, sessile, usually five pairs and an odd one, elliptic-oblong or lanceolate, acute obtuse apiculate or emarginate, quite entire or serrulate, base equal or oblique, coriaceous, nerves very inconspicuous; petiole grooved on the upper surface, sometimes faintly winged between the leaflets. *Racemes* terminal, pendulous, 6–10-flowered, peduncle short or long; pedicels slender; bracts small, linear-subulate. *Flowers* one and a half to one and three-quarters inches long. *Calyx* green, campanulate, shortly 5-lobed; lobes triangular, acute. *Corolla* scarlet; tube rather ventricose; lobes small, rounded, toothed, hairy inside towards the margins. *Stamens* four, inserted towards the base of the corolla-tube, filaments slender, hairy at the base; anthers oblong-linear, acute, those of the two longer stamens exserted, and of the shorter included. *Disk* elevated, cupular. *Ovary* flagon-shaped, glabrous, narrowed into the stout style; stigma of two oblong lobes; cells two, with two placentas inserted on the septum in each; ovules very numerous. *Capsule* 2-valved, three to four inches long, narrowly elliptic-oblong; valves coriaceous with a removable papery endocarp. *Seeds* not seen.—*J. D. H.*

Fig. 1, Flower with corolla removed; 2, corolla laid open; 3, anther; 4, ovary and disk:—*all magnified.*

TAB. 6112.

PYRUS BACCATA.

Native of Siberia, Japan, and the Himalaya Mountains.

Nat. Ord. ROSACEÆ.—Tribe POMEÆ.

Genus PYRUS, *L.* ; (*Benth. & Hook. f. Gen. Pl.*, vol. i. p. 626).

PYRUS (Malus) *baccata ;* foliis ellipticis elliptico-ovatisve acutis acuminatis v. caudato-acuminatis serrulatis glabris eglandulosis, petiolo gracili, floribus umbellatis albis, pedicellis gracilibus, calycis tubo ovoideo lobis lanceolatis intus villosis, petalis leviter concavis albis, stylis ad 5 glabris, pomo globoso apice (cicatrice calycis deciduo) late areolato.

PYRUS baccata, *Linn. Mant.*, 75; *Pall. Fl. Ross.*, vol. i. p. 23, t. 10; *DC. Prodr.*, vol. ii. p. 635; *Led. Fl. Ross.*, vol. ii. p. 97; *Loud. Arboret.*, vol. ii. p. 892; *Koch Dendrol.*, vol. i. p. 210; *Regel Gartenfl.*, vol. ii. (1862) p. 201, t. 364; *Brandis For. Flor. of N.W. India*, p. 205.

MALUS baccata, *Desf. Arb.*, vol. ii. p. 141.

This charming tree, though so long known in cultivation, has never before been well figured in this country. It has a very wide distribution; in Siberia it occurs in the eastern districts of Lake Baikal and in Dahuria; thence it passes by the Amur river north of China into Japan, whence we have numerous specimens. In the Himalaya it extends from the Indus to Kumaon, at elevations between 6000 and 11,000 feet, entering the Tibetan region of Piti; and it was gathered by Dr. Thomson and myself in the Moflong woods of the Khasia mountains, at an elevation of 6000 feet. It varies very much as to the pubescence of its parts; the Siberian and Japanese specimens being almost wholly glabrous; the Western Himalayan having more or less pubescent calyces, pedicels and petioles, and sometimes young leaves beneath; whilst those from the dry region of Piti, on the border of Tibet, are as glabrous as the Siberian; and those from the very wet region of the Khasia are the most pubescent of any. This correlation of humidity with pubescence is not unusual in the vegetable kingdom.

The figure of *P. baccata* is taken from Kew specimens,

JULY 1ST, 1874.

where the species was introduced in 1784; though whence the plant here figured came is uncertain; it will be remarked that it has a pubescent calyx-tube like the Himalayan forms.

DESCR. A small tree, with grey cracked bark, and round crown. *Leaves* one and a half to three and a half inches long, usually elliptic, acute or acuminate, finely serrate, glabrous, rarely pubescent beneath; petiole as long, very slender, glabrous, and as well as the petioles, pedicels and calyx, sometimes pubescent. *Flowers* umbelled, one and a half inches in diameter; pedicels slender. *Calyx-tube* ovoid; lobes lanceolate, deciduous. *Petals* white, rather concave, spreading. *Stamens* numerous. *Styles* 5, nearly free, glabrous or woolly at the base. *Fruit* size of a large cherry in cultivation, smaller in a native state, globose, deeply intruded at the base, with a broad apical areole, austere, scarlet and greenish yellow when ripe, endocarp almost woody in a wild state, and occupying nearly the whole fruit.—*J. D. H.*

Fig. 1, Calyx and styles :—*magnified.*

Now Ready, Part V., with 4 Coloured Plates, Royal 4to, price 5s.

ORCHIDS,
AND

How to Grow them in India & other Tropical Climates.

BY

SAMUEL JENNINGS, F.L.S., F.R.H.S.

Late Vice-President of the Agri-Horticultural Society of India.

NOW READY, Part II., 10s. 6d.

FLORA OF INDIA.

BY

DR. HOOKER, C.B., F.R.S.,

Assisted by various Botanists.

NOW READY, Vol. VI., 20s.

FLORA AUSTRALIENSIS.

A Description of the Plants of the Australian Territory. By GEORGE BENTHAM, F.R.S., assisted by BARON FERDINAND MUELLER, C.M.G., F.R.S. Vol. VI. Thymeleæ to Dioscorideæ.

NOW READY.

LAHORE TO YARKAND.

Incidents of the Route and Natural History of the Countries traversed by the Expedition of 1870, under T. D. FORSYTH, Esq., C.B. By GEORGE HENDERSON, M.D., F.L.S., F.R.G.S., Medical Officer of the Expedition, and ALLAN O. HUME, Esq., C.B., F.Z.S., Secretary to the Government of India. With 32 Coloured Plates of Birds and 6 of Plants, 26 Photographic Views of the Country, a Map of the Route, and Woodcuts. Price 42s.

In the Press and shortly to be published, in one large Volume, Royal 8vo, with numerous Coloured Plates of Natural History, Views, Map and Sections. Price 42s.

To Subscribers forwarding their Names to the Publishers before publication, 36s.

ST. HELENA:

A

Physical, Historical, and Topographical Description of the Island,

INCLUDING ITS

GEOLOGY, FAUNA, FLORA, AND METEOROLOGY.

BY

JOHN CHARLES MELLISS, C.E., F.G.S., F.L.S.

LATE COMMISSIONER OF CROWN PROPERTY, SURVEYOR AND ENGINEER OF THE COLONY.

L. REEVE AND Co., 5, Henrietta Street, Covent Garden.

DEDICATED BY SPECIAL PERMISSION TO H.R.H. THE
PRINCESS OF WALES.

NOW READY, Complete in Six Parts, 21s. each, or in One Vol., imperial folio, with 30 elaborately Coloured Plates, forming one of the most beautiful Floral Works ever published, half morocco, gilt edges, £7 7s.

A MONOGRAPH OF ODONTOGLOSSUM.

A Genus of the Vandeous section of Orchidaceous Plants. By JAMES BATEMAN, F.R.S., F.L.S., Author of "The Orchidaceæ of Mexico and Guatemala."

L. REEVE & Co., 5, Henrietta Street, Covent Garden.

NOW READY, Vol. 3, with 551 Wood Engravings, 25s.

THE NATURAL HISTORY OF PLANTS.

By Prof. H. BAILLON, P.L.S., Paris. Translated by MARCUS M. HARTOG, B.Sc., Lond., B.A., Scholar of Trinity College, Cambridge. Contents :—Menispermaceæ, Berberidaceæ, Nymphæaceæ, Papaveraceæ, Capparidaceæ, Cruciferæ, Resedaceæ, Crassulaceæ, Saxifragaceæ, Piperaceæ, Urticaceæ.

L. REEVE & Co., 5, Henrietta Street, Covent Garden.

QUADRANT HOUSE,
74, REGENT STREET, AND 7 & 9, AIR STREET, LONDON, W.

AUGUSTUS AHLBORN,

BEGS to inform the Nobility and Gentry that he receives from Paris, twice a week, all the greatest novelties and specialties in Silks, Satins, Velvets, Shawls, &c., and Costumes for morning and evening wear. Also at his establishment can be seen a charming assortment of robes for Brides and Bridesmaids, which, when selected, can be made up in a few hours. Ladies will be highly gratified by inspecting the new fashions of Quadrant House.

From *Court Journal* :—"Few dresses could compare with the one worn by the Marchioness of Bute at the State Concert at Buckingham Palace. It attracted universal attention, both by the beauty of its texture, and the exquisite taste with which it was designed. The dress consisted of a rich black silk tulle, on which were artistically embroidered groups of wild flowers, forming a most elegant toilette. The taste of the design, and the success with which it was carried out, are to be attributed to the originality and skill of Mr. AUGUSTUS AHLBORN."

Third Series.

No. 356.

VOL. XXX. AUGUST. [*Price 3s. 6d. col*^d. 2s. 6d. plain.

OR No. 1050 OF THE ENTIRE WORK.

CURTIS'S
BOTANICAL MAGAZINE,

COMPRISING

THE PLANTS OF THE ROYAL GARDENS OF KEW,

AND OF OTHER BOTANICAL ESTABLISHMENTS IN GREAT BRITAIN,
WITH SUITABLE DESCRIPTIONS;

BY

JOSEPH DALTON HOOKER, M.D., C.B., F.R.S., L.S., &c.
Director of the Royal Botanic Gardens of Kew.

Nature and Art to adorn the page combine,
And flowers exotic grace our northern clime.

LONDON:
L. REEVE & CO., 5, HENRIETTA STREET, COVENT GARDEN.
1874.

[*All rights reserved.*]

ROYAL HORTICULTURAL SOCIETY.

MEETINGS AND SHOWS IN 1874.

August   5. (Fruit and Floral Meeting.)
„   19. Do.
September 2. (Dahlias.)
October   7. { (Fruit and Floral Meeting.)
               (Fungi.)
November 11. (Fruit and Chrysanthemums.)
December 2. (Fruit and Floral Meeting.)

RE-ISSUE of the THIRD SERIES of the BOTANICAL MAGAZINE.

Now ready, Vols. I. to VIII., price 42s. each (to Subscribers for the entire Series 36s. each).

THE BOTANICAL MAGAZINE, Third Series. By Sir WILLIAM and DR. HOOKER. To be continued monthly.

Subscribers' names received by the Publishers, either for the Monthly Volume or for Sets to be delivered complete at 36s. per Volume, as soon as ready.

BOTANICAL PLATES;

OR,

PLANT PORTRAITS.

IN GREAT VARIETY, BEAUTIFULLY COLOURED, 6d. and 1s. EACH.

List of 2000 Species, one stamp.

L. REEVE & Co., 5, Henrietta Street, Covent Garden.

THE FLORAL MAGAZINE.

NEW SERIES, ENLARGED TO ROYAL QUARTO.

Figures and Descriptions of the Choicest New Flowers for the Garden, Stove, or Conservatory. Monthly, with 4 Coloured Plates, 3s. 6d. Annual Subscription, 42s.

L. REEVE & Co., 5, Henrietta Street, Covent Garden.

FLORAL PLATES,

BEAUTIFULLY COLOURED, 6d. AND 1s. EACH.

New Lists of 600 Varieties, one stamp.

L. REEVE & Co., 5, Henrietta Street, Covent Garden.

NOTICE.

THE Publishers much regret that in consequence of Dr. HOOKER's MS. having been lost in transit through the post, the Descriptions of the Plates are unavoidably delayed till next month. The names of the Plants figured are as follows:—

Tab. 6113. Crinum Moorei, *Hook. f.*—Native of South Africa.

,, 6114. Brachysema undulatum, *Ker.*—Native of Western Australia.

,, 6115. Decabelona elegans, *Decaisne.*—Native of South West Africa.

,, 6116. Kniphofia Rooperi, *Lem.*—Native of South Africa.

,, 6117. Achillea ageretifolia, *Hook. f.*—Native of Greece.

TAB. 6113.

CRINUM Moorei.

Native of South Africa.

Nat. Ord. AMARYLLIDEÆ.—Tribe AMARYLLEÆ.

Genus CRINUM, *Linn.* ; (*Herbert Amaryllid.*, p. 242).

CRINUM *Moorei*; bulbo pedali anguste ovoideo collo elongato, foliis amplis 4-poll. latis ensiformibus obtuse acuminatis striato-nervosis, scapo robusto, spathis late oblongo-lanceolatis herbaceis recurvis, pedicellis brevibus, perianthii tubo 3-pollicari, limbi 6-poll. diam. rosei segmentis late ellipticis apicibus incrassatis herbaceis, antheris flavis.

A hardy *Crinum* is a rarity in English gardens, and except the beautiful *C. capense*, I know no other but this now in open air cultivation ; and beautiful as *C. capense* is, it is far exceeded in size, foliage, and colour by the subject of the present plate.

Crinum Moorei was introduced into the Glasnevin Gardens in 1863, by a friend of Dr. Moore's, Mr. Webb, who had served on the commissariat staff of our army in South Africa, and had brought the seeds from the interior—as Dr. Moore thinks—of Natal. During the last five years the specimen from which the drawing was made has been planted in a border fronting the conservatory range at Glasnevin, without getting the slightest protection, flowering sometimes in autumn and at other times in spring. The leaves are cut up in the winter, but the bulbs are not seriously hurt, and soon recover themselves, when they push out a fresh set of their broad, peculiarly-ribbed leaves, eighteen to twenty inches long. The bulb is remarkably long, sometimes reaching eighteen inches.

A closely allied species to this is the *C. Colensoi* of Natal, which will shortly be figured, which has also broad leaves and a long bulb, but the perianth-tube is much longer, and the flower smaller, with a narrower pale limb : it has been flowered by Mr. Bull and others, and may, we hope, also prove hardy.

AUGUST 1ST, 1874.

DESCR. *Bulb* twelve to eighteen inches long, narrow ovoid contracted into a long neck. *Leaves* twelve to eighteen inches long by four broad, very numerous, erecto-patent, ensiform, with obtuse herbaceous tips, closely striated with strong nerves, deep bright green. *Scape* taller than the leaves, as thick as the thumb, green. *Spathes* six inches long, oblong-lanceolate, subacute, concave, herbaceous, reflexed. *Flowers* six to eight in a head, sessile or very shortly pedicelled. *Ovary* one inch long. *Perianth-tube* three inches long; limb four inches in diameter, very broadly campanulate, bright rose-red; segments spreading nearly from the base, broadly elliptic, with a callous green obtuse tip. *Stamen* one and a half inch long; anthers half an inch long, yellow.—*J. D. H.*

Fig. 1, Reduced figure of whole plant.

TAB. 6114.

BRACHYSEMA UNDULATUM.

Native of South-Western Australia.

Nat. Ord. LEGUMINOSÆ.—Tribe PODALYRIEÆ.

Genus BRACHYSEMA, *Br.*; (*Benth. & Hook. f. Gen. Pl.*, vol. i. p. 467).

BRACHYSEMA *undulatum*; frutex erectus, ramulis foliisque subtus sericoe-pubescentibus v. glabratis, foliis alternis subsessilibus polymorphis late oblongis ellipticis ovatis linearibusve obtusis coriaceis, floribus 1-3 pedicellatis interdum racemosis, calyce late urceolato-campanulato sericeo, lobis subacutis, vexillo cordato, alis carinæ æquilongis breviore reflexo, ovulis 15-20, legumine basi disco interiore cincto, ovoideo crustaceo piloso.

BRACHYSEMA undulatum, *Ker in Bot. Reg.* t. 642; *DC. Prodr.*, vol. ii. p. 105; *Lodd. Bot. Cab.*, t. 778; *Benth. Fl. Austral.*, vol. ii. p. 11.
B. melanopetalum, *Muell. Fragm.*, vol. iv. p. 11.
CHOROZEMA sericeum, *Smith in Trans. Linn. Soc.*, vol. ix. p. 253.
PODOLOBIUM? sericeum, *DC. Prodr.*, vol. ii. p. 103.
OXYLOBIUM? sericeum, *Benth. in Ann. Wien. Mus.*, vol. ii. p. 70.

A long known, but rare and curious greenhouse plant, remarkable for the dark violet-blue hue of the flowers, which, however, in native specimens, vary to lilac and pink. It has a wide range in Western Australia, from King George's Sound to Champion Bay, and occurs under three principal forms:—1. With broad leaves, very silky beneath, with waved margins, and usually solitary flowers; 2. With elliptic-oblong leaves, only slightly hairy beneath, hardly waved margins, and solitary flowers; this is the *B. melanopetalum* of Mueller, and that figured here; 3. With linear leaves glabrous beneath having involute margins and racemose flowers. Of these the first is the common Swan-river form, and is also found at Champion Bay; the second comes from the Tone and Don rivers, and the third from the Tone, Gordon, and Blackwood rivers.

Brachysema undulatum is a hard-wooded greenhouse shrub, requiring the same treatment as Chorozemas, &c. It was

AUGUST 1ST, 1874.

raised by Mr. Bull, with whom it flowered in April of the present year.

DESCR. A shrub, four to six feet high; young branches calyx, pedicels, and usually the leaves beneath clothed with appressed silky pubescence. *Stems* and *branches* very slender. *Leaves* one to two inches long, very shortly petioled, from linear with margins recurved to orbicular with waved margins, tip apiculate or not base, rounded, coriaceous, glabrous and dark green above, paler and usually silky beneath. *Stipules* subacute, recurved. *Flowers* three quarters of an inch long, axillary, solitary or two or three together, shortly pedicelled or racemose. *Calyx* broadly campanulate with an urceolate gibbous tube; lobes short, broad, subacute. *Petals* about twice as long as the calyx, dark purple, yellowish-green, or red; standard reflexed, cordate, shorter than the oblong obtuse wings, which equal the obtuse keel. *Ovary* hairy, many-ovuled. *Pod* short, ovoid, crustaceous.—*J. D. H.*

Fig. 1, Flowers; 2, standard; 3, wing petal; 4, keels; 5, ovary:—*all magnified*.

Tab. 6115.

DECABELONE ELEGANS.

Native of Angola.

Nat. Ord. ASCLEPIADACEÆ.—Tribe STAPELIÆ.

GEN. CHAR.—*Calyx* brevis, 5-partitus, foliolis acuminatis, glandulis herbaceis acutis interdum introrsum interpositis. *Corolla* anguste campanulata, lurida; tubo externe striis maculisque brunneo-purpureis consperso, interne pilis papillæformibus deflexis instructo; limbo 5-fido, laciniis acutis, deltoideis, paullo revolutis. *Gynostegium* imo tubo conditum. *Coronæ stamineæ* laciniæ 5, submonadelphæ, alte bifidæ, in fila gracillima apice capitato-incrassatæ attenuatæ. *Antheræ* ovatæ, obtusæ, dorso appendice ligulata incumbente; massæ pollinis horizontales, compressæ, subreniformes, funiculo appendice membranacea lineari munito. *Stigma* muticum, disciforme.—Herbæ *stapeliæformes* Africæ australis *præcipuæ tropicæ incolæ*.

DECABELONE *elegans;* caulibus ramisque stapeliæformibus angulato-costatis, costis sæpissime 6 spinosis, spinis setis lateralibus duabus erectis armatis, floribus vel e ramulorum axillis vel ad fundum inter costas ortis, coronæ stamineæ laciniis basi connatis albis, sursum in fila bina capitata atro-violacea gracillima attenuatis.

DECABELONE elegans, *Dcne. in Ann. Sc. Nat.* 5ᵃ sér. t. xiii. p. 404, pl. 2.

During the month of June of the present year this extremely interesting plant flowered for the first time in England, in the collection of J. T. Peacock, Esq., of Sudbury House, Hammersmith. A few weeks later flowers were also produced by plants in the Royal Gardens, Kew. The plate has been drawn from Mr. Peacock's specimen, a compliment which is no more than is due to the zeal and enterprise which he has shown in the cultivation of succulent plants. His plant has been grafted on a Stapelia, probably *S. Plantii.* It was obtained from Herr Pfersdorf, under the name of *Decabelone Sieberi,* but it appears to be identical with the plant described and figured by Decaisne, which was also obtained from the same cultivator, though the precise native origin was unknown.

The Kew plants were obtained direct from Angola, through the aid of Mr. Monteiro, to whom botanical science is under many obligations in elucidating the still little known Flora of that country, and was found by him at Ambriz, about three miles from the sea, in sand, near a salt marsh or "flat."

AUGUST 1ST, 1874.

The Kew Herbarium contains a specimen collected twenty years ago, at Loanda, by Dr. Welwitsch, who appears to have been disposed to constitute it a new genus, but subsequently referred it to *Huernia*. He notes that in habit it is "late cæspitosa."

The Royal Gardens, Kew, is also the fortunate possessor of specimens, both living and in spirit, as well as of drawings and analyses from His Excellency Sir Henry Barkly, Dr. Shaw, and Mrs. Barber of a second species, from Little Namaqualand. It is closely allied to *D. elegans*, the flowers being extremely similar, but the branches appear to have more numerous angles, and the two lateral setæ of the spines are more slender, and deflexed instead of erect. I am indebted to Professor Thiselton Dyer for the accompanying revised description of the genus and of this species.

DESCR. *Stems* succulent, leafless, cæspitose, four to six inches high, strongly angled, the angles furnished with patent spinous processes, each bearing two lateral erect barb-like setæ. *Flowers* produced according to Decaisne from the axils of the branchlets, but apparently also from the branches themselves between the spinous angles. *Flowers* sub-erect (the pendent habit given in the plate is due to the plant having been grafted); pedicels one-third of an inch, accompanied by one or two membranous acute bracteoles. *Calyx* 5-lobed, lobes one third to half an inch long, linear-deltoid, acuminate with a small linear herbaceous appendage arising between each pair of lobes on the inner side. *Corolla* narrowly campanulate, 5-lobed, tube externally marked with brownish-purple streaks and spots on a lurid yellow ground, internally furnished with numerous papilla-like deflexed processes and hairs; lobes deltoid, acute, slightly revolute. *Staminal-crown* 5-fid; segments connate at the base, oblong, white, deeply bifid, each tapering into two filiform capitate dark violet processes, becoming ultimately flaccid and entangled. *Anthers* broadly ovate, obtuse, with a dorsal incumbent ligulate appendage; pollen-masses horizontal, compressed, obovato-ensiform with a short funiculus furnished with a linear membranous appendage at the base, where it is attached to the stigmatic gland. *Stigma* disk-like.

Fig. 1, Spinous process from branch viewed from above (*magnified*); 2, longitudinal section through corolla (*nat. size*); 3, longitudinal section through gynostegium—the posterior pair of pollen-masses is represented displaced (*magnified*); 4, andrœcium viewed from above (*magnified*); 5, pair of pollen-masses (*magnified*).

Tab. 6116.

KNIPHOFIA Rooperi.

Native of South Africa.

Nat. Ord. LILIACEÆ.—Tribe ALOINEÆ.

Genus KNIPHOFIA, *Mœnch.; (Endl. Gen. Plant.,* p. 143).

KNIPHOFIA *Rooperi;* acaulis, foliis elongato-ensiformibus 1¾ poll. latis alte carinatis tenuiter cartilagineo-serrulatis, scapo valido, bracteis caulinis paucis brevibus e basi lata semiamplexicauli subulatis, racemo ovoideo-oblongo, perianthii recti 1½ pollicaris lobis brevibus obtusis, bracteolis latis, filamentis demum exsertis.

KNIPHOFIA Rooperi, *Moore in Gard. Comp.,* vol. i. p. 113 (Tritoma); *Lemaire Jard. Fleur.,* t. 362; *Baker in Journ. Linn. Soc.,* vol. xi. p. 363.

This is very nearly allied to the well known *K. aloides,* (*K. Uvaria,* Tab. 4816, *Tritoma Uvaria,* Tab. 758), and may perhaps prove to be a late flowering variety of it; in which opinion I am strengthened by Mr. Baker, who has monographed the genus in the Linnean Journal. The chief character by which this was distinguished, namely, the included stamens, does not hold good, as the plate shows; better ones may be found in the paler, less curved flower, in the form of the bracteoles, and in broad rich glaucous leaves.

Of the fourteen species enumerated by Mr. Baker, seven have been figured in this country from cultivated specimens; namely, 1. *K. aloides,* mentioned above, which was introduced in 1707, according to the Hortus Kewensis, and probably much earlier into Europe, as it is mentioned in Stapel's Theophrastus as "Iris Uvaria promontorii Bonæ spei;" 2. *K. præcox,* Baker, (Saund. Refug. Bot. t. 168); 3. *K. Burchellii,* Kunth (Bot. Reg. t. 1745); 4. *K. pumila* (Tab. nost. 764); 5. *K. sarmentosa* (Tab. nost. 744, *Iris media*); 6. *K. caulescens* (Tab. nost. 5946), and the present plant. All are probably hardy, and require protection only during very severe winters; indeed, it is to the latter cause alone that can be attributed the loss during half a century of so conspicuous and easily grown a plant as K. *aloides,* which reappeared in cultivation

AUGUST 1ST, 1874.

not very many years ago. *K. Rooperi* is a native of British Kaffraria, whence it was sent to England by Capt. Rooper, whose name it bears. The specimen here figured flowered with Mr. Wilson Saunders in November of last year.

DESCR. Two feet high. *Stem* none. *Leaves* eighteen inches long by one and three-quarters broad, ensiform, gradually acuminate, deeply keeled, at the back dark green, not glaucous, margin serrulate. *Scape* very stout, a foot long; bracts few, short, membranous, subulate from a broad semiamplexicaul base. *Spike* six to eight inches long, ovoid-oblong. *Flowers* densely crowded, about one and a half inches long, orange-red, becoming yellow with age; bracteoles broadly lanceolate, acute. *Stamens* at length exserted.—*J. D. H.*

Tab. 6117.

ACHILLEA AGERATIFOLIA.

Native of Greece.

Nat. Ord. COMPOSITÆ.—Tribe ANTHEMIDEÆ.

Genus ACHILLEA, *Linn.*; (*Benth. & Hook. f. Gen. Plant.*, vol. i. p. 419).

ACHILLEA (Ptarmica) *ageratifolia*; perennis, cæspitosa, tota niveo-tomentosa, foliis radicalibus confertis patenti-recurvis anguste lingulato-lanceolatis apices versus dilatatis obtusis crenato-serratis crenis interdum biseriatis, caulinis sæpe basi dilatatis et pectinato-lobulatis, capitulis solitariis amplis, involucri hemispherici squamis oblongis, paleis oblongo-lanceolatis apice scariosis laceris, ligulis 2-seriatis late oblongis 3-crenatis.

ACHILLEA ageratifolia, *Benth. in Benth. et Hook. f. Gen. Pl.*, vol. i, p. 420.

ANTHEMIS ageratifolia, *Sibth. Prodr. Flor. Græc.*, vol. ii. p. 191; *Fl. Græc.*, t. 888; *DC. Prodr.*, vol. vi. p. 12.

? A. Aizoon, *Griseb. Fl. Runel. et Byth.*, vol. ii. p. 210; *Walp. Rep.*, vol. vi. p. 187.

? A. Aizoides, *Boiss. et Orphan. mss.*

This charming little plant is a native of the mountains of Greece, and was first detected (in Crete?) by Sibthorp, and it has since been gathered on the mainland, by Prof. Orphanides, of Athens, in the middle region of Mount Olympus in Thessaly, at an elevation of 5-7000 feet. A very nearly allied plant, most probably a variety, is the *Anthemis Aizoon*, of Griesbach, a native of the mountains of Macedonia, which has also been gathered in the upper regions of Mount Parnassus, at an elevation of 6-7000 feet, by Prof. Orphanides, and named by him and M. Boissier *Anthemis Aizoides*; it differs in the smaller size, shorter more spathulate leaves, which show no signs of double crenature. Both are obviously species of Achillea, having compressed achenes, quite different from those of Anthemis, of which they have more the habit.

Lindley, in the ninth volume of Sibthorp's Flora Græca, observes correctly, that this is not the *Lepidophorum repandum*,

AUGUST 1ST, 1874.

as suspected by De Candolle; he describes it from very imperfect specimens, which he found in Sibthorp's herbarium, along with *Gnaphalium luteo-album*, from Crete. It appears to me so unlikely that these plants should have grown together, that I suspect some confusion of habitat, and that Sibthorp did not collect his plant in Crete, where no one has found it, but on Mount Olympus, which he visited, and where he could not well have missed finding it.

The Royal Gardens are indebted to Mr. Niven, of the Hull Botanic Gardens, for living plants which flowered at Kew in May.

DESCR. Covered with white soft tomentum. *Roots* woody. *Stems* many, short, tufted. *Leaves* spreading, recurved; radical one to one and a half inches long, linear-lingulate, obtuse, pectinately crenate, crenatures often in two series; cauline linear, obtuse or subacute, base at times dilated and pectinate. *Flowering-stems* six to ten inches high. *Heads* solitary, one to one and a quarter inches diameter, white with a pale yellow disk. *Involucre* hemispherical; scales with broad very obtuse scarious margins; paleæ linear-lanceolate with scarious toothed tips. *Ray-flowers* in two series, broadly oblong, with three toothed tips and a winged tube. *Disk-flowers* with a winged lower half of the tube. *Achenes* obovate, flattened, winged.—*J. D. H.*

Fig. 1, Leaf; 2, flower of ray; 3, do. of disk:—*all magnified.*

Now Ready, Part VI., with 4 Coloured Plates, Royal 4to, price 5s.

ORCHIDS,
AND
How to Grow them in India & other Tropical Climates.
BY
SAMUEL JENNINGS, F.L.S., F.R.H.S.
Late Vice-President of the Agri-Horticultural Society of India.

NOW READY, Part II., 10s. 6d.

FLORA OF INDIA.
BY
DR. HOOKER, C.B., F.R.S.
Assisted by various Botanists.

NOW READY, Vol. VI., 20s.

FLORA AUSTRALIENSIS.
A Description of the Plants of the Australian Territory. By GEORGE BENTHAM, F.R.S., assisted by BARON FERDINAND MUELLER, C.M.G., F.R.S. Vol. VI. Thymeleæ to Dioscorideæ.

NOW READY.

LAHORE TO YARKAND.
Incidents of the Route and Natural History of the Countries traversed by the Expedition of 1870, under T. D. FORSYTH, Esq., C.B. By GEORGE HENDERSON, M.D., F.L.S., F.R.G.S., Medical Officer of the Expedition, and ALLAN O. HUME, Esq., C.B., F.Z.S., Secretary to the Government of India. With 32 Coloured Plates of Birds and 6 of Plants, 26 Photographic Views of the Country, a Map of the Route, and Woodcuts. Price 42s.

In the Press and shortly to be published, in one large Volume, Royal 8vo, with numerous Coloured Plates of Natural History, Views, Map and Sections. Price 42s.

To Subscribers forwarding their Names to the Publishers before publication, 36s.

ST. HELENA:
A
Physical, Historical, and Topographical Description of the Island,
INCLUDING ITS
GEOLOGY, FAUNA, FLORA, AND METEOROLOGY.
BY
JOHN CHARLES MELLISS, C.E., F.G.S., F.L.S.
LATE COMMISSIONER OF CROWN PROPERTY, SURVEYOR AND ENGINEER OF THE COLONY.

L. REEVE & Co., 5, Henrietta Street, Covent Garden.

DEDICATED BY SPECIAL PERMISSION TO H.R.H. THE
PRINCESS OF WALES.

NOW READY, Complete in Six Parts, 21s. each, or in One Vol., imperial folio, with 30 elaborately Coloured Plates, forming one of the most beautiful Floral Works ever published, half morocco, gilt edges, £7 7s.

A MONOGRAPH OF ODONTOGLOSSUM.

A Genus of the Vandeous section of Orchidaceous Plants. By JAMES BATEMAN, F.R.S., F L.S., Author of "The Orchidaceæ of Mexico and Guatemala."

L. REEVE & Co., 5, Henrietta Street, Covent Garden.

NOW READY, Vol. 3, with 551 Wood Engravings, 25s.

THE NATURAL HISTORY OF PLANTS.

By Prof. H. BAILLON, P.L.S., Paris. Translated by MARCUS M. HARTOG, B.Sc., Lond., B.A., Scholar of Trinity College, Cambridge. Contents:—Menispermaceæ, Berberidaceæ, Nymphæaceæ, Papaveraceæ, Capparidaceæ, Cruciferæ, Resedaceæ, Crassulaceæ, Saxifragaceæ, Piperaceæ, Urticaceæ.

L. REEVE & Co., 5, Henrietta Street, Covent Garden.

QUADRANT HOUSE,

74, REGENT STREET, AND 7 & 9, AIR STREET, LONDON, W.

AUGUSTUS AHLBORN,

Begs to inform the Nobility and Gentry that he receives from Paris, twice a week, all the greatest novelties and specialties in Silks, Satins, Velvets, Shawls, &c., and Costumes for morning and evening wear. Also at his establishment can be seen a charming assortment of robes for Brides and Bridesmaids, which, when selected, can be made up in a few hours. Ladies will be highly gratified by inspecting the new fashions of Quadrant House.

From the *Court Journal*:—"Few dresses could compare with the one worn by the Marchioness of Bute at the State Concert at Buckingham Palace. It attracted universal attention, both by the beauty of its texture, and the exquisite taste with which it was designed. The dress consisted of a rich black silk tulle, on which were artistically embroidered groups of wild flowers, forming a most elegant toilette. The taste of the design, and the success with which it was carried out, are to be attributed to the originality and skill of Mr. AUGUSTUS AHLBORN."

Third Series.

No. 357.

VOL. XXX. SEPTEMBER. [*Price 3s. 6d. col^{d.} 2s. 6d. plain.*

OR No. 1051 OF THE ENTIRE WORK.

CURTIS'S
BOTANICAL MAGAZINE,

COMPRISING

THE PLANTS OF THE ROYAL GARDENS OF KEW,

AND OF OTHER BOTANICAL ESTABLISHMENTS IN GREAT BRITAIN,
WITH SUITABLE DESCRIPTIONS;

BY

JOSEPH DALTON HOOKER, M.D., C.B., F.R.S., L.S., &c.

Director of the Royal Botanic Gardens of Kew.

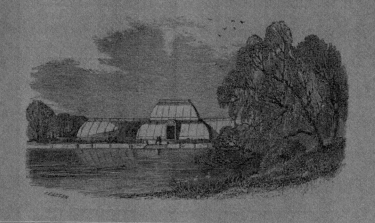

Nature and Art to adorn the page combine,
And flowers exotic grace our northern clime.

LONDON:
L. REEVE & CO., 5, HENRIETTA STREET, COVENT GARDEN.
1874.

[*All rights reserved.*]

ROYAL HORTICULTURAL SOCIETY.

MEETINGS AND SHOWS IN 1874.

September 2. (Dahlias.)
October 7. { (Fruit and Floral Meeting.) (Fungi.) }
November 11. (Fruit and Chrysanthemums.)
December 2. (Fruit and Floral Meeting.)

RE-ISSUE of the THIRD SERIES of the BOTANICAL MAGAZINE.

Now ready, Vols. I. to IX., price 42s. each (to Subscribers for the entire Series 36s. each).

THE BOTANICAL MAGAZINE, Third Series. By Sir WILLIAM and DR. HOOKER. To be continued monthly.

Subscribers' names received by the Publishers, either for the Monthly Volume or for Sets to be delivered complete at 36s. per Volume, as soon as ready.

BOTANICAL PLATES;
OR,
PLANT PORTRAITS.

IN GREAT VARIETY, BEAUTIFULLY COLOURED, 6d. and 1s. EACH.

List of 2000 Species, one stamp.

L. REEVE & Co., 5, Henrietta Street, Covent Garden.

THE FLORAL MAGAZINE.

NEW SERIES, ENLARGED TO ROYAL QUARTO.

Figures and Descriptions of the Choicest New Flowers for the Garden, Stove, or Conservatory. Monthly, with 4 Coloured Plates, 3s. 6d. Annual Subscription, 42s.

L. REEVE & Co., 5, Henrietta Street, Covent Garden.

FLORAL PLATES,

BEAUTIFULLY COLOURED, 6d. AND 1s. EACH.

New Lists of 600 Varieties, one stamp.

L. REEVE & Co., 5, Henrietta Street, Covent Garden.

TAB. 6118.

IRIS TECTORUM.

Native of Japan.

Nat. Ord. IRIDACEÆ.—Tribe IRIDEÆ.

Genus IRIS, *Linn.*; (*Endl. Gen. Plant.*, p. 266).

IRIS *tectorum;* rhizomate crasso, caule elato, foliis pedalibus $\frac{3}{4}$-$1\frac{1}{4}$ poll. latis læte viridibus, scapo subcompresso foliis subæquante, spathis oblongis acutis herbaceis 3-valvibus, floribus 3-4 poll. diam., pedicello ovario æquilongo, perianthii tubo crassiusculo ore 6-gluniiloso, segmentis crispato-undulatis subæqualibus exterioribus lilacinis maculatis obovato-rotundatis reflexis ungue albido venis violaceis, crista laciniata, interioribus unicoloribus, filamentis complanatis, stigmatibus ligulatis superne dilatatis, segmentis grossæ dentatis.

IRIS tectorum, *Maxim. Diagn. brev. Pl. Nov. Jap. Dec.*, viii. p. 563; *Regel Garten-Fl.*, vol. xxi. p. 65, t. 716.

I. tomiolopha, *Hance in Trimen Jour. Bot. N.S.*, vol. i. p. 229.

I. cristata, *Miq. Pro. Fl. Jap.*, p. 305, *non Ait.*

Although the plant here figured came from Whampoa in China, where it was cultivated by Dr. Hance, Her Britannic Majesty's Vice-Consul at that port, there can be no question but that it is the Japanese *Iris tectorum* of Maximovicz, which grows in fields about Yokohama in Japan, and is likewise cultivated by the Japanese. It differs from Maximovicz's description, but not from native specimens, in having three spathes, which are acute or acuminate—characters which (with some others of foliage that are very variable) induced Dr. Hance to publish it as a new species, under the name of *tomiolopha*, in allusion to its cut crest. On the other hand it differs from Dr. Hance's description in the spreading inner perianth-segments, a character probably due to cultivation, as it occurs in the splendid *Iris Kæmpferi* var. *Hendersoni*, lately exhibited in the Royal Horticultural Society by Messrs. Henderson, and which is unquestionably a form of *I. lævigata*, with a spreading perianth. With the North American *I. cristata*, to which it was referred by the late

SEPTEMBER, 1874.

Professor Miquel, it has no near affinity, but it has with the Himalyan *I. decora*, Wall., of Nepal.

I am indebted to Mr. Bull for the specimen here figured, which was raised from seeds sent by Dr. Hance from his garden in Whampoa, and which flowered in April, 1874.

DESCR. *Rootstock* creeping, tuberous, annulate. *Leaves* about a foot long, by three-quarters to one and a quarter inches broad, ensiform, scarcely glaucous. *Scape* nearly terete, about as long as the leaves. *Flowers* three to four inches in diameter. *Spathes* three, herbaceous, green, erect, longer than the perianth-tube, acute. *Pedicel* about as long as the ovary. *Perianth-tube* one inch long; outer segments one and a half inches broad, obovate, margin crisped and waved, pale lilac streaked with violet; claw half as long as the limb, white; crest running up the claw and half the limb, half an inch deep, white and lilac, deeply laciniate; inner segments rather narrower than the outer, spreading, pale lilac, claw short. *Stigmas* half as long as the perianth-segments recurved, lilac, segments coarsely toothed.—*J. D. H.*

TAB. 6119.

BOLBOPHYLLUM Dayanum.

Native of Tenasserim.

Nat. Ord. ORCHIDEÆ.—Tribe DENDROBIEÆ.

Genus BOLBOPHYLLUM, *Thouars;* (*Lindl. Gen. & Sp. Orchid.*, p. 47).

BOLBOPHYLLUM *Dayanum;* rhizomate crasso cylindrico repente, pseudobulbis globoso-ovoideis sulcatis, folio oblongo obtuso crasse coriaceo-carnoso enervi costa subtus prominente floribus 2-3, 1 poll. diam. ringentibus in racemulo abbreviato subsessilibus, sepalis ovatis obtusis saturate viridibus purpureo-maculatis longe ciliatis, petalis multo minoribus lineari-oblongis acutis purpureis ciliatis, labelli parvi pallide viridis vix unguiculati lobis lateralibus parvis auriculæformibus crenulatis, terminali late oblongo obtuso crenato-dentato, disco cristis 3-elevatis centralibus crenatis ornato, et utrinque intra cristas et margines seriebus 3 spinularum aucto, columna apice dentata.

BOLBOPHYLLUM Dayanum, *Reichb. f. in Gard. Chron.*, 1865, 434; *Xenia Orchidacea*, p. 128, t. 144.

This singularly coloured species of Bolbophyllum was introduced by Mr. Day from Moulmein, and published in the *Gardeners' Chronicle* for 1865, by Professor Reichenbach; and it has now again been sent to England by our old friend Mr. Parish, who has returned to the scene of his clerical and botanical labours, which he has resumed with his wonted energy and former success. It is very much to be wished that these might culminate in a general work from his pen on the Orchids of the Tenasserim provinces, that have proved such a mine of wealth in these plants, and for which Mr. Parish's special acquirements admirably suit him. As it is, Orchidology is falling into a hopeless condition, and but for the generous assistance of Professor Reichenbach, both cultivators and botanists would be very badly off indeed. A synopsis of the genera and species, or even a classified catalogue of these, with synonyms and habitats and references to publications, would be a boon to Botany and Horticulture. Of Bolbophyllum alone no less than eighty-four species were brought together by Prof. Reichenbach in the sixth volume of Walper's

"Annales," which professes to bring the subject down to 1855; and who but Dr. Reichenbach knows how many have been published since—and where?

DESCR. *Rhizome* creeping, stout, as thick as a goosequill, smooth, woody, annulate. *Pseudobulbs* globose-ovoid, deeply channelled, with rounded ridges. *Leaf* shortly petioled, three to four inches long, by one quarter to two inches broad, thickly coriaceous, almost fleshy, oblong, tip obtuse, recurved; midrib stout and convex below, deep green above, purplish beneath, nerveless. *Flowers* about three in a very shortly peduncled raceme or umbel from the base of the pseudobulb, one inch in diameter. *Ovary* curved, pedicel short, stout. *Sepals* spreading, ovate, obtuse, ciliate, with long spreading hairs, yellow-green, with six rows of dark purple spots. *Petals* about one quarter the size of the sepals, spreading, linear-oblong, obtuse, ciliate, purple with green edges. *Lip* very small, shortly clawed, pale purple edged with green, lateral lobes small oblong crenulate; midlobe broadly oblong obtuse crenate, with three longitudinal crests in the disk, of which the lateral are raised towards the base into flat erect crenate plates; there are also three rows of spinous processes on each side of the lip.—*J. D. H.*

Fig. 1, Flower; 2, sepal; 3, column and lip; 4, lip:—*all magnified.*

TAB. 6120.

CINNAMODENDRON CORTICOSUM.

Native of the West Indies.

Nat. Ord. CANELLACEÆ.

Genus CINNAMODENDRON, *Endl.; (Benth. & Hook. f. Gen. Pl.*, vol. i. p. 121).

CINNAMODENDRON *corticosum*; glaberrimum, foliis anguste elliptico-v. obovato-oblongis -lanceolatisve sæpe gibbis v. inæquilateris obtusis v. subacutis marginibus recurvis basi rotundatis acutisve, cymis parvis axillaribus, sepalis 5 ovato-oblongis apice recurvis rotundatis, petalis 4–5 erectis oblongis, antheris 16–20.

CINNAMODENDRON corticosum, *Miers Contrib. to Bot.*, vol. i. p. 121, t. 24; *Griseb. Fl. Brit. W.-Ind.*, 109.

A well known West Indian tree, as the Mountain Cinnamon of Jamaica, or *Canella* bark of that island and St. Thomas, but not the true Brazilian plant of that name, which is its solitary congener, the *C. axillare* of Endlicher. These two very distinct trees were indeed long confounded together, and their bark is still imported under the same name of *Canella*, and employed largely as an aromatic stimulant to purgatives and tonics, being reputed to be well adapted for debilitated stomachs. The Caribs (ancient natives of the Antilles) and the negroes of the present day employ it is a condiment. As an aromatic, Pereira says that it ranks between cinnamon and cloves. Mr. Hanbury informs me that the bark was exported during the last century as "Winter's bark" and is still found in the market; as also that it is probably the "Wild Cinnamon tree of Sloane, commonly but falsely called Cortex Winteranus," though the tree that he figures Phil. Trans. xvii. 465, (1693) is certainly *Canella alba*. It is a local plant growing in Jamaica only in mountain woods of the parishes of St. Thomas in the Vale and St. John.

In the following description I have followed the view of the nature of the outer floral whorls adopted in the Genera Plantarum, though more disposed to regard the outermost three organs as sepals, and the innermost four or five as staminal appendages.

DESCR. A small or large tree, fifty feet high, branched from the base, glabrous throughout; branches terete; bark aromatic. *Leaves* alternate, shortly petioled, four to five inches long, oblong-lanceolate, narrowed at both ends, subacute or obtuse, base rounded or acute, often very unequal-sided, one side bulging out above the middle, coriaceous, covered with pellucid dots, margins recurved, nerves 10–12 pairs, very slender, reticulations delicate; petioles one quarter inch to one-third inch. *Cymes* axillary, very shortly peduncled, 6–8-flowered, quite glabrous. *Flowers* shortly pedicelled, one-third inch in diameter, orange-red, pedicels short, bracts at their bases deciduous. *Bracteoles* (or sepals) three, orbicular, green, ciliolate. *Sepals* (or petals) five, erect with recurved rounded tips, oblong-ovate, red. *Petals* (or staminal scales), four or five, erect, linear-oblong, unequal. *Staminal column* cylindric, 5-lobed at the apex; anthers sixteen to twenty, linear. *Ovary* seated in a cupular disk, obtusely 3-gonous, ovoid-oblong, obscurely contracted into a columnar style whose rounded apex is divided into three to five minute stigmatic lobes. *Berry* ovoid, half an inch long, many-seeded. *Seeds* ovoid, testa brown shining, albumen fleshy and oily; embryo linear.—*J. D. H.*

Fig. 1, Flower; 2, the same with bracts and three sepals removed; 3 staminal column; 4, portion of do.; 5, disk and ovary; 6, transverse section of ovary:—*all magnified.*

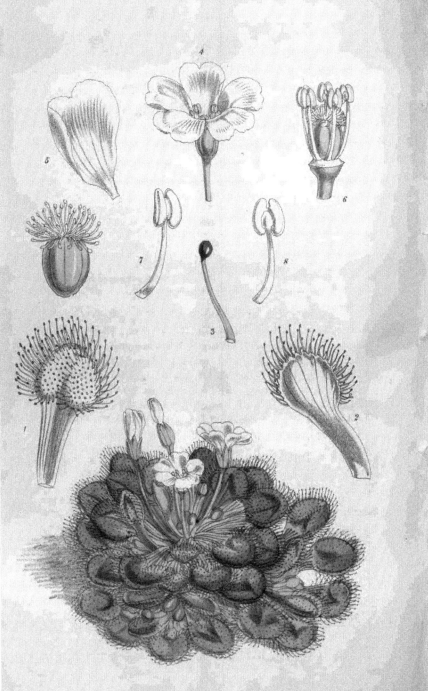

Tab. 6121.

DROSERA Whittakerii.

Native of South Australia.

Nat. Ord. DROSERACEÆ.

Genus DROSERA, *Linn.*; (*Benth. & Hook. f. Gen. Pl.*, vol. i. p. 662).

DROSERA *Whittakerii*; tubere globoso corticato cortice chartaceo fragili nitido, rhizomate erecto valido subterraneo, foliis rosulatis spathulato-obovatis viridibus crassiusculis, pilis glandulosis rubris elongatis, scapis brevibus 1-floris sepalisque elongatis viridibus eglandulosis, petalis obcordato-cuneatis albis, ovario globoso, stylis in laciniis capillaribus capitellatis ad basim fissis.

DROSERA Whittakerii, *Hook. Ic. Pl.*, t. 375; *Planch. in Ann. Sc. Nat.*, ser. 3, vol. ix. p. 202; *Benth. Flor. Austral.*, vol. i. p. 462.

This charming little plant was sent to the Royal Botanic Gardens of Edinburgh by Mr. W. A. Mitchell, formerly an *employé* in that establishment, where it was flowered by Mr. McNab in July last, and sent up to Kew for figuring, with a description by my friend Dr. Balfour, who observed that the sepals were reflexed, and the flowers an inch in diameter when well grown and expanded, a statement fully borne out by the dried specimens. The glandular hairs on the leaf are in all respects like those of *D. longifolia*, and act precisely in the same manner on being brought into contact with insects; the leaf itself, however, does not become concave, but retains the remarkable convexity of surface of each half.

Drosera Whittakerii is a common Victorian and South Australian plant, and belongs to a group of very closely allied species, including *D. bulbosa, zonata* and *rosulata*, all having tuberous roots, attaining a considerable size in the first named of these. Such sorts indicate a totally different kind of treatment to what answers in cultivation for *D. rotundifolia* and its allies; it indicates a resting season, preceded by one for the ripening of the bulb, and followed by a growing one in due course. The same remark applies to almost all terrestrial Australian Orchids, objects of inconceivable beauty and interest, but which have never been successfully kept in this country.

DESCR. *Leaves* rosulate, very numerous, densely crowded, one to one and a quarter inches long, one-half to three-quarters inch wide at the broadest part, obovate-spathulate, the petiolar part broad green, the blade tumid on the face on either side the mesial line, and studded with long red-brown glandular hairs, rather fleshy. *Scapes* several, about equalling the leaves, slender, erect and one-flowered, quite glabrous and not glandular. *Flower* one-half inch to one inch in diameter. *Sepals* oblong, obtuse, green, glabrous, eglandular. *Petals* obcordate-cuneate, white. *Stamens* quite hypogynous; anther-cells separated by the connective. *Ovary* globose; styles split at the base into filiform capitate white filaments.—J. D. H.

Fig. 1 and 2, Leaves; 3, glandular hair of do.; 4, flower; 5, petal; 6, top of pedicel, stamen, and pistil; 7 and 8, stamens; 9, ovary:—*all magnified.*

Tab. 6122.

PENTSTEMON humilis.

Native of the Rocky Mountains.

Nat. Ord. Scrophularineæ.—Tribe Cheloneæ.

Genus Pentstemon, *L'Her.*; (*Benth. in DC. Prodr.*, vol. x. p. 320).

Pentstemon *humilis*; glaberrimus, foliis radicalibus anguste lineari-lanceolatis ellipticis v. elliptico-spathulatis petiolatis integerrimis v. obscure crenatis acuminatis caulinis oblongis linearibusve, floribus racemosis breviter gracile pedicellatis, calycis lobis recurvis lanceolatis ciliatis, corolla semi-pollicari, tubo lente curvo modice inflato, limbo cæruleo 2-labiato, labii superioris lobis breviter oblongis inferioris late obovatis obtusis, filamentis glabris, antherarum loculis divaricatis, stylo piloso.

Pentstemon humilis, *Nutt. in Herb. Acad. Philad. ex A. Gray in Proc. Amer. Acad. Arts and Sc.*, October, 1862, p. 69; *S. Watson Bot. 40th Parallel*, pp. 220 and 454.

The charming little plant here figured differs very much in stature and foliage from the indigenous specimens preserved in the Kew Herbarium, which are eight inches to a foot high, more robust, and have elliptic-ovate radical leaves, oblong-spathulate cauline ones, and flowers two-thirds of an inch long. All these, however, are differences of degree only, and I quite expect that older specimens of the cultivated plant will assume the stature and probably the foliage of the native ones. Add to these points the known variability of the species of Pentstemon, and that there is no other species to which the present bears any resemblance (except the foliage to the otherwise very different *P. Hallii*), and no doubt is left in my mind as to the identification of this with *P. humilis*.

Pentstemon humilis was one of the indefatigable Nuttall's discoveries in the Rocky mountains, and it has since been gathered by the naturalists attached to various American and English Government-exploring expeditions, amongst others, by Dr. Lyall, of the Oregon Boundary Commission, who collected it at 7000 feet above the sea, between Fort Colville and the Rocky mountains, in 1867. The plant here represented was sent for figuring by Messrs. Backhouse, of York, who flowered it in June last.

DESCR. *Root* perennial, bearing many short branches. *Leaves* chiefly radical, from linear-lanceolate to elliptic-oval, obtuse acute or acuminate, coriaceous, quite entire, nerveless, glabrous. *Flowering-stems* six to twelve inches high, erect, with two or more pairs of linear-lanceolate or oblong leaves. *Panicle* of three or more whorls of shortly pedicelled flowers; bracts herbaceous, ovate-lanceolate, acute. *Flower* half to two-thirds of an inch long, horizontal or drooping. *Calyx* small, glandular, pubescent or ciliate; segments subulate-lanceolate, recurved. *Corolla-tube* moderately inflated, pale reddish-purple; throat open, not folded; limb 2-lipped, bright blue purple; upper lip of two shortly oblong rounded lobes; under of three shortly broadly obovate lobes. *Filaments* glabrous; anther-cells divergent. *Style* hairy.—*J. D. H.*

Fig. 1, Flower; 2, base of corolla-tube, stamens and style:—*all magnified.*

Tab. 6123.

BRODIÆA VOLUBILIS.

Native of California.

Nat. Ord. LILIACEÆ.—Tribe MILLEÆ.

Genus BRODIÆA, *Smith*; (*Baker in Journ. Linn. Soc.*, vol. xi. p. 375).

BRODIÆA (Stropholirion) *volubilis*; cormo globoso, foliis synanthiis carnoso-herbaceis 1–1½ pedalibus ¼–⅓ poll. latis, scapo volubili prælongo, spathis 4–5 oblongo-lanceolatis, umbellis 15–30 floris, perianthio campanulato-infundibuliformi tubo subventricoso, segmentis suberectis obtusis, antheris sessilibus alatis, staminodiis ligulatis, ovario breviter stipitato.

BRODIÆA volubilis, *Baker in Journ. Linn. Soc.*, vol. xi. p. 377.

STROPHOLIRION californicum, *Torrey in Bot. Whipple Exped.*, p. 149, t. 23; *Benth. Plant. Hartweg*, 339.

RUPALLEYA volubilis, *Morière in Bull. Soc. Linn. Norm.*, viii. *cum ic ex Bull. Soc. Bot. France*, vol. xi. Bibl. 25.

DICHELOSTEMMA californica, *Wood in Proc. Philad. Acad.*, 1867, 173.

It is not surprising that so remarkable a plant as this should have been erected into a genus; or that, considering the chaotic state of North American descriptive botany, it should have had two made on purpose for it; or that it should in fact have been referred by name to three other genera before Mr. Baker, in his revision of the *Liliaceæ*, reduced it to its proper position as *Brodiæa*, reserving for it, however, as a sectional name, Torrey's generic one of *Stropholirion*. For the justice of this view I would refer to our plate of the floral structure of *Brodiæa multiflora* (Tab. 5989), where it will be seen, that except by the twining scape, *Stropholirion* differs from that genus in no important particular.

Brodiæa volubilis was discovered by Hartweg in the Sacramento mountains, California, in 1846, and has since been found by various collectors in Sonora and other places. The scape sometimes attains twelve feet in length.

The plant figured was raised and sent for figuring by Mr. Thompson, of Ipswich, in July of the present year.

SEPTEMBER, 1874.

DESCR. *Corm* the size of a walnut. *Leaves* a foot long, narrowly linear-lanceolate, acuminate, trigonous, acutely keeled at the back, channelled in front, very pale green. *Scape* four to twelve feet long, twining amongst the branches of bushes, one quarter inch in diameter, green varied with pink. *Umbel* large, three to four inches in diameter, of very many (twelve to twenty) pedicelled rosy flowers; pedicels a quarter to one inch long; spathes four or five, oblong-lanceolate, concave, shorter than the rosy pedicels, tipped with green. *Flower* three-quarters of an inch long. *Perianth* between campanulate and funnel-shaped; tube 5-lobed, tumid; segments erect, ovate, obtuse. *Anthers* three, opposite the inner perianth lobes; adnate to and winged by the broad filament behind it. *Staminodes* ligulate, notched, pubescent. *Ovary* shortly stipitate, ellipsoid 3-gonous.—*J. D. H.*

Fig. 1, Flower; 2, vertical section of do.; 3, staminode:—*all magnified.*

Now Ready, Part VII., with 4 Coloured Plates, Royal 4to, price 5s.
ORCHIDS,
AND
How to Grow them in India & other Tropical Climates.
BY
SAMUEL JENNINGS, F.L.S., F.R.H.S.
Late Vice-President of the Agri-Horticultural Society of India.

NOW READY, Part II., 10s. 6d.
FLORA OF INDIA.
BY
DR. HOOKER, C.B., F.R.S.
Assisted by various Botanists.

NOW READY, Vol. VI., 20s.
FLORA AUSTRALIENSIS.
A Description of the Plants of the Australian Territory. By GEORGE BENTHAM, F.R.S., assisted by BARON FERDINAND MUELLER, C.M.G., F.R.S. Vol. VI. Thymeleæ to Dioscorideæ.

NOW READY.
LAHORE TO YARKAND.
Incidents of the Route and Natural History of the Countries traversed by the Expedition of 1870, under T. D. FORSYTH, Esq., C.B. By GEORGE HENDERSON, M.D., F.L.S., F.R.G.S., Medical Officer of the Expedition, and ALLAN O. HUME, Esq., C.B., F.Z.S., Secretary to the Government of India. With 32 Coloured Plates of Birds and 6 of Plants, 26 Photographic Views of the Country, a Map of the Route, and Woodcuts. Price 42s.

In the Press and shortly to be published, in one large Volume, Royal 8vo, with numerous Coloured Plates of Natural History, Views, Map and Sections. Price 42s.

To Subscribers forwarding their Names to the Publishers before publication, 36s.

ST. HELENA:
A
Physical, Historical, and Topographical Description of the Island,
INCLUDING ITS
GEOLOGY, FAUNA, FLORA, AND METEOROLOGY.
BY
JOHN CHARLES MELLISS, C.E., F.G.S., F.L.S.
LATE COMMISSIONER OF CROWN PROPERTY, SURVEYOR AND ENGINEER OF THE COLONY.

L. REEVE & Co., 5, Henrietta Street, Covent Garden.

DEDICATED BY SPECIAL PERMISSION TO H.R.H. THE
PRINCESS OF WALES.

NOW READY, Complete in Six Parts, 21s. each, or in One Vol., imperial folio,
with 30 elaborately Coloured Plates, forming one of the most beautiful
Floral Works ever published, half morocco, gilt edges, £7 7s.

A MONOGRAPH OF ODONTOGLOSSUM.

A Genus of the Vandeous section of Orchidaceous Plants. By JAMES BATEMAN,
F.R.S., F.L.S., Author of "The Orchidaceæ of Mexico and Guatemala."

L. REEVE & Co., 5, Henrietta Street, Covent Garden.

NOW READY, Vol. 3, with 551 Wood Engravings, 25s.

THE NATURAL HISTORY OF PLANTS.

By Prof. H. BAILLON, P.L.S., Paris. Translated by MARCUS M. HARTOG, B.Sc., Lond., B.A., Scholar of Trinity College, Cambridge. Contents:—Menispermaceæ, Berberidaceæ, Nymphæaceæ, Papaveraceæ, Capparidaceæ, Cruciferæ, Resedaceæ, Crassulaceæ, Saxifragaceæ, Piperaceæ, Urticaceæ.

L. REEVE & Co., 5, Henrietta Street, Covent Garden.

QUADRANT HOUSE,

74, REGENT STREET, AND 7 & 9, AIR STREET, LONDON, W.

AUGUSTUS AHLBORN,

Begs to inform the Nobility and Gentry that he receives from Paris, twice a week, all the greatest novelties and specialties in Silks, Satins, Velvets, Shawls, &c., and Costumes for morning and evening wear. Also at his establishment can be seen a charming assortment of robes for Brides and Bridesmaids, which, when selected, can be made up in a few hours. Ladies will be highly gratified by inspecting the new fashions at Quadrant House.

From the *Court Journal*:—"Few dresses could compare with the one worn by the Marchioness of Bute at the State Concert at Buckingham Palace. It attracted universal attention, both by the beauty of its texture, and the exquisite taste with which it was designed. The dress consisted of a rich black silk tulle, on which were artistically embroidered groups of wild flowers, forming a most elegant toilette. The taste of the design, and the success with which it was carried out, are to be attributed to the originality and skill of Mr. AUGUSTUS AHLBORN."

Third Series.

No. 358.

VOL. XXX. OCTOBER. [*Price 3s. 6d. col*^d *2s. 6d. plain.*]

OR No. 1052 OF THE ENTIRE WORK.

CURTIS'S
BOTANICAL MAGAZINE,

COMPRISING

THE PLANTS OF THE ROYAL GARDENS OF KEW,

AND OF OTHER BOTANICAL ESTABLISHMENTS IN GREAT BRITAIN,
WITH SUITABLE DESCRIPTIONS;

BY

JOSEPH DALTON HOOKER, M.D., C.B., F.R.S., L.S., &c.

Director of the Royal Botanic Gardens of Kew.

Nature and Art to adorn the page combine,
And flowers exotic grace our northern clime.

LONDON:
L. REEVE & CO., 5, HENRIETTA STREET, COVENT GARDEN.
1874.

[*All rights reserved.*]

ROYAL HORTICULTURAL SOCIETY.

MEETINGS AND SHOWS IN 1874.

October 7. { (Fruit and Floral Meeting.)
{ (Fungi.)
November 11. (Fruit and Chrysanthemums.)
December 2. (Fruit and Floral Meeting.)

RE-ISSUE of the THIRD SERIES of the BOTANICAL MAGAZINE.

Now ready, Vols. I. to X., price 42s. each (to Subscribers for the entire Series 36s. each).

THE BOTANICAL MAGAZINE, Third Series. By Sir WILLIAM and DR. HOOKER. To be continued monthly.

Subscribers' names received by the Publishers, either for the Monthly Volume or for Sets to be delivered complete at 36s. per Volume, as soon as ready.

BOTANICAL PLATES;

OR,

PLANT PORTRAITS.

IN GREAT VARIETY, BEAUTIFULLY COLOURED, 6d. and 1s. EACH.

List of 2000 Species, one stamp.

L. REEVE & Co., 5, Henrietta Street, Covent Garden.

THE FLORAL MAGAZINE.

NEW SERIES, ENLARGED TO ROYAL QUARTO.

Figures and Descriptions of the Choicest New Flowers for the Garden, Stove, or Conservatory. Monthly, with 4 Coloured Plates, 3s. 6d. Annual Subscription, 42s.

L. REEVE & Co., 5, Henrietta Street, Covent Garden.

FLORAL PLATES,

BEAUTIFULLY COLOURED, 6d. AND 1s. EACH.

New Lists of 600 Varieties, one stamp.

L. REEVE & Co., 5, Henrietta Street, Covent Garden.

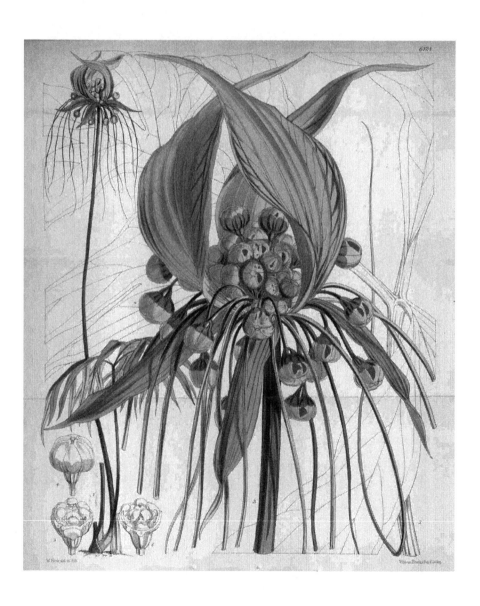

TAB. 6124.

TACCA ARTOCARPIFOLIA.

Native of Madagascar and Johanna.

Nat. Ord. TACCACEÆ.

Genus TACCA, *Forst.*; (*Endl. Gen. Pl.*, vol. i. p. 159).

TACCA *artocarpifolia*; elata, foliis amplis trisectis, segmentis petiolulatis 1½–2 pedalibus pinnatifidis, laciniis pedalibus 1–1½ poll. latis sensim acuminatis, scapo elongato, involucro 6–7-phyllo, foliolis 4–8-pollicaribus caudato-acuminatis exterioribus deflexis angustioribus, interioribus laterioribus conniventibus erectis elliptico-lanceolatis, omnibus integerrimis v. exteriore 3-fido, pedicellis sterilibus pedalibus filiformibus, ovario turbinato alte 6-carinato, perianthio globoso.

TACCA artocarpifolia, *Seemann Flor. Vit.*, p. 101, *in note.*

This very remarkable plant is a congener of the *Ataccia cristata* of this work (Tab. 4589), the genus *Ataccia* being now regarded as a synonym of *Tacca*, and of *T. integrifolia* (Tab. 1488). It is a native of Madagascar and Johanna Islands, whence there are excellent specimens in the Hookerian Herbarium, from Mr. Justice Blackburn, Dr. Lyall, and W. T. Gerrard. Its nearest ally is the well known *T. pinnatifida*, which, though one of the most widely cultivated and most useful plants in the Pacific Islands, has never yet been figured in any English botanical work; nor, as far as we know, ever been introduced into this country. The tubers of *T. pinnatifida* afford the South-Sea arrowroot, said to be the best of all in cases of dysentery, and its starch is a favourite article of diet in the shape of puddings and cakes.

The *T. artocarpifolia* has a tuberous root, and is, no doubt, as full of starch and as wholesome as *T. pinnatifida*. It flowered in the Royal Gardens in May of the present year, from roots received from Mr. Wilson Saunders in 1872.

DESCR. *Root* tuberous. *Leaves* about three; petiole two feet long, stout, erect, cylindric, nearly as thick as a goosequill, brown; base curved with thick sheathing wings; blade two to three feet in diameter, trisect; segments stalked, pinnatifid but not to the base; pinnules three to four pairs, one and a half feet long, by one to one and a half inches broad, spreading, gradually narrowed into long acuminate points; midrib stout, lateral nerves very slender, elongate, and running parallel to the margins. *Scape* five to six feet high, as thick as the little finger, brown. *Involucre* of six to seven leaves; outer narrow-lanceolate, five to six inches long, deflexed; inner erect, incurved, conniving, elliptic-lanceolate, all strongly nerved, green, caudate-acuminate. *Flowers* very numerous, pedicels one to three inches long; sterile pendulous ones filiform, ten to twelve inches long, brown, grooved on one face. *Ovary* turbinate, with six strong keeled ribs, brown. *Perianth* glabrous, brown at the base, the rest green; segments broadly ovate, conniving, coriaceous. *Stamens* sessile. *Stigma* 3-lobed, lobes convex. *Fruit* six inches long, ellipsoid-oblong, 6-ribbed.—*J. D. H.*

Fig. 1, *Reduced* figures of leafing and flowering states; 2, portion of leaf and 3, inflorescence :—of the *natural size;* 4, flowers; 5, the same with the perianth-segments removed; 6, the same with 3 stamens removed showing the stigma :—*all magnified.*

TAB. 6125.

POGONIA DISCOLOR.

Native of Java.

Nat. Ord. ORCHIDEÆ.—Tribe ARETHUSEÆ.

Genus POGONIA, *Juss.*; (*Lindl. Gen. et Sp. Orchid.*, 413).

POGONIA *discolor*; folio breviter petiolato rotundato-cordato-multinervio discolori supra et subtus ad nervos rufo-setoso, scapo subbifloro bracteis spathaceis occulto, sepalis petalisque consimilibus linearibus acuminatis pallide fusco-viridibus, labello obcordato-2-lobo albido.

P. discolor, *Blume Mus. Bot. Lugd.-Bat.*, vol. i. p. 32; *Coll. Orchid. Archip. Ind. et Jap.*, 152, t. 57, f. 1; *Miquel Fl. Ind.-Bat.*, vol. iii. p. 716.

RIPHOSTEMON discolor, *Blume Flor. Jav. Præf.*, vol. vi.; *Lindl. Gen. et Sp. Orchid.*, p. 453.

CORDYLA discolor, *Blume Bijd.*, p. 417.

The species of *Pogonia* have usually little to recommend them for horticultural purposes; but to this there are exceptions, especially amongst the Indian species, some of which that have been cultivated at Kew present, like that here figured, beautifully coloured and marked leaves that persist for many weeks, and attract the attention of the most ordinary observer. All have tuberous roots, often formed at the end of subterranean cylindric fibres. It is not easy so to manage their culture as that the leaves, flowers, and new tubers should be successfully formed, and upon this their continuance under cultivation depends. The present is closely allied to the common *P. plicata* of Bengal, which has a rose-coloured lip. Blume describes the lip of *P. discolor* as entire, but it is retuse in his drawing, and distinctly 2-lobed in our specimen.

Pogonia discolor is a native of damp forests in the mountain region of Western Java, where it flowers in November. The specimen here figured was flowered by Mr. Bull, in February last, and the leaf was fully formed in the following June.

OCTOBER 1ST, 1874.

DESCR. *Root* of small spherical tubers. *Leaf* solitary, three to five inches in diameter, nearly horizontal, orbicular-cordate, cuspidate, convex, membranous; upper surface dark rufous green, often with paler blue-green blotches between the nerves, clothed with long rufous bristles which are disposed in lines along the principal and secondary nerves; under surface pale dull purple, with bristles on the principal nerves only; nerves radiating from the top of the petiole, eleven to sixteen, rufous above; petiole half to one inch high. *Scape* solitary, two to three inches long, clothed with the loose pale dirty-green or purplish bracteal sheaths. *Flowers* in pairs, one and a half inches in diameter. *Ovary* turbinate, sharply angled, short, glabrous. *Sepals* and *petals* equal and similar, spreading, three-quarters of an inch long, linear, acuminate, dirty grey-green. *Lip* as long as the petals, white, with a green disk, convolute for nearly its whole length; limb obcordate, 2-lobed, erose; disk crested. *Column* erect, clavate, white. *Pollen-masses* oblong.—*J. D. H.*

Fig. 1, Side, and 2, front view of ovary and lip; 3 column; 4 and 5, pollen masses:—*all magnified.*

Tab. 6126.

LILIUM MACULATUM.

Native of Japan and N.E. Asia.

Nat. Ord. LILIACEÆ.—Tribe TULIPEÆ.

Genus LILIUM, *Linn.; (Baker in Gard. Chron.* 1871; *Journ. Hort. Soc.*
N.S. vol. iv. p. 39).

LILIUM (Martagon) *maculatum;* glaberrimum, bulbo solitario squamis fusi-
formibus apice articulatis, foliorum verticillis 1-3, foliis in verticillo
4-20 lineari-lanceolatis ellipticisve obtusis v. obtuse acuminatis, racemo
1-12 flore, floribus 2-3 poll. diametr. cernuis, perianthio aurantiaco
basi late campanulato, foliolis patenti-recurvis oblongis obtusis basim
versus rima nuda nectarifera instructis medium versus punctis majusculis atro-purpureis notatis, capsula pyriformi vertice depresso abrupte
in pedicellum brevem attenuata.

L. maculatum, *Thunb. in Mem. Acad. Petersb.,* vol. iii. t. 5; *Baker in Journ. Hort. Soc. Lond.,* N.S. vol. iv. p. 45.

L. avenaceum, *Fischer—F. Schmidt Flor. Sachalin ined. ex Maximovicz in Regel Gartenfl.,* vol. xiv. (1865) 290, t. 485.

L. Martagon, *Ledeb. Fl. Ross.,* vol. iv. p. 149, *quoad plantam Kamtscha-
ticam.*

A native of Kamtschatka, the Kurile and Sachalin Islands, Japan, and South Eastern Manchuria, whence it was introduced into the Russian Imperial Botanical Gardens, and we have dried specimens from Sitcha on the American coast.

According to Maximovicz (in *Gartenflora*), there are two varieties of it: one, with a scented orange-yellow flower, which is that figured here, and which came from Victoria Sound; the other, with red inodorous flowers, is found in Japan and Kamtschatka, is figured by Regel in the *Gartenflora*. The leaves of both varieties vary extremely, both as to the number of whorls, the number in each whorl, and in length and breadth.

I am indebted to G. F. Wilson, F.H.S., of Weybridge Heath, for the specimen figured, the bulb of which he purchased at a sale of Japan Lilies, in London, and which flowered in June of the present year. I have also seen a specimen flowered by Mr. William Saunders, F.H.S, in 1871.

OCTOBER 1ST, 1874.

DESCR. *Bulb* from the size of a large nut to that of a walnut; scales fleshy, fusiform. *Stem* two to three feet high, slender, bearing one to three whorls of four to twenty leaves at various heights above the ground, but always far above it. *Leaves* narrow when many, broad when few, sessile, three to six inches long by one half to one and a half inches broad, linear- or elliptic-lanceolate or elliptic, obtuse or obtusely acuminate, with three to five principal nerves, dark green above, paler beneath; upper leaves alternate scattered, much smaller. *Flowers* usually four to six (one to twenty), irregularly racemose; peduncles two three inches long; bracts broad, green, leafy. *Perianth* two three inches in diameter, campanulate at the base, orange-yellow or red, with black rounded or oblong spots about the middle of the segments which are spreading and recurved with obtuse thickened points. *Filaments* much shorter than the perianth; anthers linear, yellow. *Ovary* oblong; style short, clavate, stigma hemispherical. *Capsule* (according to Regel) pyriform.—*J. D. H.*

Fig. 1, Ovary and style;—*magnified.*

Tab. 6127.

SCORZONERA undulata.

Native of Algeria and Marocco.

Nat. Ord. Compositæ.—Tribe Cinchonaceæ.

Genus Scorzonera, *Linn.*; (*Benth. & Hook. f. Gen. Pl.*, vol. ii. p. 531).

Scorzonera *undulata;* caule erecto gracili elongato ramoso glabro, foliis glabris radicalibus pedalibus anguste elongato-lanceolatis attenuato-acuminatis vix undulatis, caulinis e basi lata sessile subulato-attenuatis, capitulis 2–2½ poll. diam., involucri bracteis cano-tomentellis exterioribus late ovatis acutis apicibus recurvis, interioribus ter longioribus ovato-oblongis anguste marginatis, ligulis roseo-purpureis, corollæ tubo gracili apicem versus barbellato, achænio obconico costato costis crenulatis, pappi setis inæqualibus infra medium plumosis, 5–6 ceteris longioribus robustioribus supra medium scabridis.

S. undulata, *Vahl. Symb.*, vol. ii. p. 85; *Desfont. Flor. Atlant.*, vol. ii. p. 219; *DC. Prodr.*, vol. vii. p. 117 (*in part*).

That this is the true *Scorzonera undulata* of Vahl and of Desfontaine I have little doubt, though it differs somewhat from the description of the former in its flat leaves and more branched habit, and altogether from specimens in the Herbarium of the Greek plant with which Desfontaine and De Candolle confound the Barbary one. Vahl describes it as common throughout the region of Tunis, and Desfontaine has found it in sandy places. In many respects, and especially in habit, it approaches *S. purpurea*, Linn., which has smaller bracts and very narrow leaves; and also *S. hispanica*, which differs chiefly, if not only, in its yellow flowers.

The specimen here figured was brought by Mr. Maw from Algeria, where he recognised it as apparently the same with a plant we found between Tangiers and Tetuan in Marocco. It flowered in July, and had a very handsome appearance.

Descr. *Root* thick, fusiform. *Stem* one to two feet, slender, branched, glabrous or with scanty white tomentum towards the ends of the branches. *Radical leaves* nearly a foot long, narrowly linear-lanceolate, quite entire glabrous, three-quarters of an inch in diameter at the broadest part, gradually narrowed into a rather long petiole and to the acuminate apex, bright green, with a yellow midrib; margins scarcely waved; *cauline-leaves* three to six inches long, gradually narrowed from a broad sessile base to a very fine point, erect, keeled to the midrib, margins even. *Peduncles* slender, green or purplish. *Heads* two to two and a half inches in diameter, rose-purplish; involucre cylindric, tumid at its base, hoary with a white down; outer scales broadly ovate, with a recurved acuminate apex, green with pale margins; inner twice or thrice as long, linear-oblong, with broad scarious margins. *Flowers* numerous. *Corolla-tube* slender, bearded below the mouth; limb as long, linear, 5-toothed. *Achenes* (unripe) fimbriate, grooved, the ridges crenulate; pappus-hairs about twenty, all plumose halfway up, five or six stronger than the rest and scabrid from above the middle to the apex.—*J. D. H.*

Fig. 1, Flowers; 2, unripe achene and pappus:—*both magnified.*

Tab. 6128.

CITRUS Aurantium var. japonica.

Cultivated in China and Japan. The Kumquat.

Nat. Ord. Rutaceæ.—Tribe Aurantieæ.

Genus Citrus, *Linn.*; (*Benth. & Hook. f. Gen. Pl.*, v. i. p. 305).

Citrus *Aurantium* var. *japonica;* fruticosa, spinosa v. inermis, ramis angulatis, petiolis cuneato-alatis v. lineari-cuneatis, foliis elliptico- v. oblongo-lanceolatis obtuse acuminatis apices versus crenulatis, floribus axillaribus solitariis fasciculatisve albis, fructibus globosis v. ellipsoideo-oblongis 4–6-locularibus, cortice granulato.

Citrus japonica, *Thunb. Flor. Japan.*, 292; *Ic.* t. 15; *DC. Prodr.*, vol. i. 540; *Sieb. & Zucc. Fl. Jap.*, vol. i. p. 35, t. 15; *Fortune in Hort. Soc. Journ.*, N.S. vol. ii. p. 46; *Gard. Chron.* 1874, 336, cum ic. xylog.; *Miq. Prolus. Fl. Jap.* p. 15.

C. Margarita, *Lour. Fl. Coch.*, p. 570; *DC.* l.c.

Kinkan, *Kœmpf. Amœn. Exot.*, vol. v. p. 801.

Subvar. inermis.—C. inermis, *Roxb. Fl. Ind.*, vol. iii. p. 393. C. madurensis, *Lour. Fl. Coch.*, 570; *DC. Prodr.*, l.c.; *Rumph. Herb. Amb.*, vol. ii. p. 110, t. xxxi.

This well-known ingredient in Chinese sweetmeats has never previously been figured from cultivated specimens in Europe, though long known from Kæmpfer's description and plate. According to Siebold, it is nowhere found wild in Japan; this author says that, in common with all other species and varieties of *Citrus*, it has been introduced into the Island from China or India; also that it is extensively cultivated under two varieties, one with globose, the other with oval fruit, which latter is rare. He adds, that the agreeable acid of the juice, flavoured by the aroma of the rind, renders the fruit very pleasant, but that it yields only a transient refreshment, for it leaves a burning after-taste in the mouth.

A magnificent fruiting specimen of this interesting shrub was exhibited by Mr. Bateman at the Horticultural Society in 1867, from which the accompanying drawing was taken. It belonged to the unarmed variety, and is far more luxuriant, both as to foliage and fruit, than the dried specimen, or those figured by Siebold and Zuccarini. As regards the cultivation of the *Kumquat*, Mr. Fortune, who introduced it, says in his paper published in the *Journal of the Horticultural Society*, quoted above, that it requires in summer plenty of water at a temperature of 80° to 100°, and a high atmospheric heat continued into autumn; whilst in winter it should be kept cool and rather dry, for it will then bear 10° and even 15° of frost. It succeeds well grafted on *Limonia trifoliata*.

DESCR. A shrub or small tree, four to six feet high. *Branchlets* green, glabrous, compressed, trigonous. *Leaves* biennial; petiole one-third to one-half inch long, narrowly cuneate or almost linear; blade three to five inches long, elliptic or oblong-lanceolate, narrowed at both ends, obtuse, crenate above the middle. *Flowers* one to three, axillary, fascicled, three-fourths of an inch to one inch in diameter, white; peduncles glabrous. *Calyx* short, five-lobed, glabrous, green, segments broad. *Petals* oblong, subacute. *Stamens* twenty or fewer, irregularly connate into bundles. *Disk* thick. *Ovary* 4–6-celled. *Fruit* two-thirds to one inch in diameter, globose or shortly ellipsoid, bright orange-yellow, 4–6-celled; rind thick, minutely tuberculate; pulp watery, sweet and acidulous. *Seeds* few, like those of the common orange, but much smaller.—*J. D. H.*

Fig. 1, Calyx and Stamens—*magnified*; 2, transverse section of the fruit of *the natural size*.

Now Ready, Part VIII., with 4 Coloured Plates, Royal 4to, price 6s.

ORCHIDS,
AND
How to Grow them in India & other Tropical Climates.

BY

SAMUEL JENNINGS, F.L.S., F.R.H.S.

Late Vice-President of the Agri-Horticultural Society of India.

NOW READY, Part II., 10s. 6d.

FLORA OF INDIA.

BY

DR. HOOKER, C.B., F.R.S.

Assisted by various Botanists.

NOW READY, Vol. VI., 20s.

FLORA AUSTRALIENSIS.

A Description of the Plants of the Australian Territory. By GEORGE BENTHAM, F.R.S., assisted by BARON FERDINAND MUELLER, C.M.G., F.R.S. Vol. VI. Thymeleæ to Dioscorideæ.

NOW READY.

LAHORE TO YARKAND.

Incidents of the Route and Natural History of the Countries traversed by the Expedition of 1870, under T. D. FORSYTH, Esq., C.B. By GEORGE HENDERSON, M.D., F.L.S., F.R.G.S., Medical Officer of the Expedition, and ALLAN O. HUME, Esq., C.B., F.Z.S., Secretary to the Government of India. With 32 Coloured Plates of Birds and 6 of Plants, 26 Photographic Views of the Country, a Map of the Route, and Woodcuts. Price 42s.

In the Press and shortly to be published, in one large Volume, Royal 8vo, with numerous Coloured Plates of Natural History, Views, Map and Sections. Price 42s.

To Subscribers forwarding their Names to the Publishers before publication, 36s.

ST. HELENA:

A

Physical, Historical, and Topographical Description of the Island,

INCLUDING ITS

GEOLOGY, FAUNA, FLORA, AND METEOROLOGY.

BY

JOHN CHARLES MELLISS, C.E., F.G.S., F.L.S.

LATE COMMISSIONER OF CROWN PROPERTY, SURVEYOR AND ENGINEER OF THE COLONY.

L. REEVE & Co., 5, HENRIETTA STREET, COVENT GARDEN.

DEDICATED BY SPECIAL PERMISSION TO H.R.H. THE PRINCESS OF WALES.

NOW READY, Complete in Six Parts, 21s. each, or in One Vol., imperial folio, with 30 elaborately Coloured Plates, forming one of the most beautiful Floral Works ever published, half morocco, gilt edges, £7 7s.

A MONOGRAPH OF ODONTOGLOSSUM.

A Genus of the Vandeous section of Orchidaceous Plants. By JAMES BATEMAN, F.R.S., F L.S., Author of "The Orchidaceæ of Mexico and Guatemala."

L. REEVE & Co., 5, Henrietta Street, Covent Garden.

NOW READY, Vol. 3, with 551 Wood Engravings, 25s.

THE NATURAL HISTORY OF PLANTS.

By Prof. H. BAILLON, P.L.S., Paris. Translated by MARCUS M. HARTOG, B.Sc., Lond., B.A., Scholar of Trinity College, Cambridge. Contents :—Menispermaceæ, Berberidaceæ, Nymphæaceæ, Papaveraceæ, Capparidaceæ, Cruciferæ, Resedaceæ, Crassulaceæ, Saxifragaceæ, Piperaceæ, Urticaceæ.

L. REEVE & Co., 5, Henrietta Street, Covent Garden.

QUADRANT HOUSE,

74, REGENT STREET, AND 7 & 9, AIR STREET, LONDON, W.

AUGUSTUS AHLBORN,

BEGS to inform the Nobility and Gentry that he receives from Paris, twice a week, all the greatest novelties and specialties in Silks, Satins, Velvets, Shawls, &c., and Costumes for morning and evening wear. Also at his establishment can be seen a charming assortment of robes for Brides and Bridesmaids, which, when selected, can be made up in a few hours. Ladies will be highly gratified by inspecting the new fashions of Quadrant House.

From the *Court Journal* :— " Few dresses could compare with the one worn by the Marchioness of Bute at the State Concert at Buckingham Palace. It attracted universal attention, both by the beauty of its texture, and the exquisite taste with which it was designed. The dress consisted of a rich black silk tulle, on which were artistically embroidered groups of wild flowers, forming a most elegant toilette. The taste of the design, and the success with which it was carried out, are to be attributed to the originality and skill of Mr. AUGUSTUS AHLBORN."

Third Series.

No. 359.

VOL. XXX. NOVEMBER. [Price 3s. 6d. col^{d.} 2s. 6d. plain.

OR No. 1053 OF THE ENTIRE WORK.

CURTIS'S
BOTANICAL MAGAZINE,

COMPRISING

THE PLANTS OF THE ROYAL GARDENS OF KEW,

AND OF OTHER BOTANICAL ESTABLISHMENTS IN GREAT BRITAIN, WITH SUITABLE DESCRIPTIONS;

BY

JOSEPH DALTON HOOKER, M.D., C.B., F.R.S., L.S., &c.

Director of the Royal Botanic Gardens of Kew.

Nature and Art to adorn the page combine,
And flowers exotic grace our northern clime.

LONDON:
L. REEVE & CO., 5, HENRIETTA STREET, COVENT GARDEN.
1874.

[All rights reserved.]

ROYAL HORTICULTURAL SOCIETY.

MEETINGS AND SHOWS IN 1874.

November 11. (Fruit and Chrysanthemums.)
December 2. (Fruit and Floral Meeting.)

RE-ISSUE of the THIRD SERIES of the BOTANICAL MAGAZINE.

Now ready, Vols. I. to XI., price 42s. each (to Subscribers for the entire Series 36s. each).

THE BOTANICAL MAGAZINE, Third Series. By Sir WILLIAM and DR. HOOKER. To be continued monthly.

Subscribers' names received by the Publishers, either for the Monthly Volume or for Sets to be delivered complete at 36s. per Volume, as soon as ready.

L. REEVE & Co., 5, Henrietta Street, Covent Garden.

BOTANICAL PLATES;
OR,
PLANT PORTRAITS.

IN GREAT VARIETY, BEAUTIFULLY COLOURED, 6d. and 1s. EACH.

List of 2000 Species, one stamp.

L. REEVE & Co., 5, Henrietta Street, Covent Garden.

THE FLORAL MAGAZINE.

NEW SERIES, ENLARGED TO ROYAL QUARTO.

Figures and Descriptions of the Choicest New Flowers for the Garden, Stove, or Conservatory. Monthly, with 4 Coloured Plates, 3s. 6d. Annual Subscription, 42s.

L. REEVE & Co., 5, Henrietta Street, Covent Garden.

FLORAL PLATES,

BEAUTIFULLY COLOURED, 6d. AND 1s. EACH.

New Lists of 600 Varieties, one stamp.

L. REEVE & Co., 5, Henrietta Street, Covent Garden.

TAB. 6129.

PASSIFLORA (TACSONIA) MANICATA.

Native of New Grenada and Peru.

Nat. Ord. PASSIFLORACEÆ.—Tribe PASSIFLOREÆ.

Genus PASSIFLORA, *Linn.*; (*Benth. & Hook. f. Gen. Plant.*, vol. i. p. 810).

PASSIFLORA (Tacsonia) *manicata;* caule flexuoso subangulato, foliis 4-pollicaribus coriaceis 3-lobis serratis supra glabris subtus pubescentibus, lobis ovatis acutis intermedio producto, petiolo pollicari 3–4-gianduloso, stipulis dimidiato-ovatis falcatis grosse dentatis, pedunculo petiolo duplo longiore, bracteis pollicaribus ellipticis acutis serrulatis liberis v. connatis, floribus coccineis, perianthii tubo ½–1-pollicari basi dilatato limbo 4-pollicari.

TACSONIA manicata, *Juss. in Ann. Muss.*, vol. vi. p. 393, t. 59, f. 2; *Lindl. & Paxt. Fl. Gard.*, vol. i. t. 26; *DC. Prodr.*, vol. iii. p. 334; *Humb. Bonpl. & Kunth Nov. Gen.*, vol. ii. p. 139; *Masters in Mart. Flor. Bras.*, vol. xiii. pars i. p. 541.

This lovely plant has been for many years cultivated in England, though not so extensively as it deserves, having had the reputation of not flowering freely. It was introduced previous to 1850 by the Horticultural Society, through its collector Hartweg, who found it in hedges near Loxa in Peru, where, indeed, it was discovered by Humboldt and Bonpland half a century previously. It is also a native of the Andes of Equador and New Grenada, where it was found by Purdie on the arid plains of Suta Marchan, and is there called *Ruruba de Seneno*. A similar undescribed species, or perhaps a variety of this, with white flowers, was gathered by Pearce at Puquina (in Peru ?), at an elevation of 10,000 feet.

I am indebted for the accompanying drawing to Mr. E. J. Smith, of Coalport, in whose conservatory the plant flowered in July last.

I regret not being able to follow Dr. Masters in retaining the genus *Tacsonia*, as is done in his very admirable Monograph of South American Passifloræ, in Martius's "Flora Brasiliensis;" the only character hitherto adopted being the

NOVEMBER 1ST, 1874.

comparative length of the perianth-tube of *Tacsonia*, which is shorter in this species than in various Brazilian plants universally referred to *Passiflora*. Could genera be limited by geographical distribution, *Tacsonia* would (as Dr. Masters indicates) be defined as being confined, as far as is hitherto known, to the Andes of South America, whilst the *Passifloras* are spread over the warm regions of the whole American continent, and are found also in Asia and Australia.

DESCR. *Stem* climbing, nearly terete, and as well as the petioles leaves beneath stipules bracts and perianth externally finely pubescent. *Leaves* about four inches long, coriaceous, 3-lobed to about the middle, finely serrate; lobes broadly oblong, obtuse or subacute, dark green above, pale beneath, base rounded truncate or subacute; petiole about one inch long, with three to four glands. *Stipules* one inch in diameter, dimidiate-ovate, deeply toothed, convex. *Peduncle* longer than the petiole. *Bracts* at some distance from the calyx, elliptic-ovate, acute, serrate, pubescent, free or united from the base upwards, sometimes for half their length. *Perianth-tube* about half an inch long, base inflated and 10-lobed; limb four inches in diameter, vivid scarlet; corona double; outer, at the mouth of the tube, of many series of short blue hairs, the inner row of which connives around the column; inner, at the top of the inflated base of the perianth, formed of a sigmoidly-inflexed membrane. *Styles* free.—*J. D. H.*

Fig. 1, Vertical section of perianth tube:—*somewhat magnified.*

TAB. 6130.

CERINTHE GYMNANDRA.

Native of Italy, Algeria, and Marocco.

Nat. Ord. BORRAGINEÆ.—Tribe CERINTHEÆ.

Genus CERINTHE, *Linn.; (A. DC. Prodr.*, vol. 10, p. 2).

CERINTHE *gymnandra;* annua, glabra, caule subflexuoso simplici v. ramoso, foliis oblongis ovato-oblongisve apice sphacelatis obtuse arcuatis v. rotundatis basi auriculato 2-lobis, supra remote verruculosis subtus glabris, calycis foliolis lineari-oblongis erectis ciliatis, corolla curva infra medium clavato-inflato, lobis 5 triangulari-subulatis reflexis, antherarum caudis setaceis apicibus exsertis.

CERINTHE gymnandra, *Gaspar., Rendei dell Acad. Soc. Real Borbon di Nap.*, vol. i. p. 72, *ex Reichb. Ic. Fl. Germ.*, vol. xviii. p. 36, t. 1297; *Willkom. & Lange Fl. Hisp.*, vol. ii. p. 512.

A very rare European plant, hitherto found, as far as I am aware, only near Naples, whence I have seen specimens collected by Heldreich; it is however common in some parts of Western Algeria, as at Oran and Blidah, growing in sandy places, and in Marocco. Though hardly different from *C. major* (Tab. nost. 333), as pointed out by Willkomm and Lange, it is a very beautiful form of the genus, well worthy of cultivation, but unfortunately annual. One of its most striking characters is the discoloration of the tips of the leaves; these in all our specimens are of a fine pale glaucous blue, except at the very end, which is pale greenish-yellow, bounded towards the midrib by a dull dark purple band; thus the colouring of the leaf-tip is a repetition of that of the flower, and gives a bright appearance to the whole plant. From the above-quoted figure of *Cerinthe major* in this Magazine, the present differs in the yellow tubular terminal portion of the corolla, the narrower sepals not cordate at the base, and foliage; but little dependence can be placed on these characteristics in so variable a genus.

Our specimens were raised from seed sent by Messrs. Haage and Schmidt, and flowered in July.

NOVEMBER 1ST, 1874.

DESCR. Annual, very variable in size and stature, glabrous, except the calyx. *Stem* six to twelve inches high, usually ascending from a short base, simple or branched above, rather stout, quite smooth, pale yellow-green. *Leaves* one to four inches long, glaucous, usually ovate-oblong and somewhat contracted in the middle, rounded or obtusely pointed; base 2-lobed, with two deep rounded auricles; upper surface with scattered small warts, only visible in the dry state, under surface quite smooth; tips always discoloured, yellow-green with a purple band beneath it, strongly contrasting with the glaucous blue of the rest of the leaf; nerves faint. *Floral leaves* large, distichous, imbricate, enclosing and almost concealing the flowers. *Flowers* shortly pedicelled, nearly one inch long. *Calyx* half the length of the corolla; leaflets linear-oblong, acute, ciliate, with a purple band below the tip. *Corolla* curved, lower parts rather inflated, subclavate, deep red-purple except at the base; upper part cylindric, yellow; lobes short, triangular, subulate, yellow, sharply reflexed. *Filaments* short; anthers slender, with exserted subulate purple tips, the cells caudate at the base.—*J. D. H.*

Fig. 1, Flower; 2, corolla laid open; 3, stamen; 4, disk, ovary and style:— *all magnified.*

TAB. 6131.

MELALEUCA WILSONI.

Native of South Australia.

Nat. Ord. MYRTACEÆ.—Tribe LEPTOSPERMEÆ.

Genus MELALEUCA, *Linn.*; (*Benth. Fl. Austral.*, vol. iii. p. 123).

MELALEUCA *Wilsoni*; ramulis junioribus exceptis glaberrima, foliis oppositis confertis patulis v. in ramulis junioribus imbricatis subulato-lanceolatis acutis v. acuminatis subtus convexis obscure 1-3-nerviis, floribus solitariis v. fasciculatis sparsis v. spicatis, calycis tubo ovoideo glabro basi rotundato lobis ovato-lanceolatis subacutis tubum æquantibus, petalis calycis lobis duplo longioribus ellipticis concavis, staminum phalangiis erecto-patentibus ½—pollicaribus spathulatis, filamentis 15–20.

MELALEUCA Wilsoni, *F. Muell. Fragm.*, vol. ii. p. 124, t. 15; *Benth. Fl. Austral.*, v. iii. p. 134.

This is one of that large class of hard-wooded Australian plants which, if properly cultivated, would ornament our conservatories and greenhouses at seasons when little else worth looking at meets the eye, but which have almost throughout the country succumbed to the treatment they have received—namely, of watering in season and out of season. The genus to which it belongs contains just one hundred species, scattered over all parts of Australia, amongst which are some of the most brilliant-coloured plants of that gay Flora. The present is essentially a dry country species, inhabiting the desert of the Tattiave country, Port Lincoln, &c., in South Australia, as also the country around Lake Hindmarsh in the colony of Victoria. It was raised at Kew from seeds sent by Baron Muller from the Melbourne Botanic Garden, when he was director of that rich botanical establishment; and was named by him after Mr. Charles Wilson, through whose aid, he states, this very beautiful species was discovered.

DESCR. A slender shrub, glabrous, except the puberulous young branches. *Leaves* close set, spreading, those on the young branches imbricate, one-third to three-quarters of an inch long, subulate-lanceolate, rigid, quite entire, concave below, obscurely 3-nerved, obtuse or pungent. *Flowers* crowded, rarely solitary, sometimes forming cylindric spikes, rarely solitary, sessile. *Bracts* imbricate, membranous, equalling the calyx-tube. *Calyx* green, tube one-twelfth of an inch long, ovoid, rounded at the base; lobes rather shorter, erect, ovate, subacute. *Petals* about twice as long as the calyx-lobes, elliptic, subacute, concave, erect. *Bundles of stamens* erect, then spreading, half an inch long, bright rose-red; claw linear-spathulate; filaments fifteen to twenty. *Fruit* 3-valved—*J. D. H.*

Fig. 1, Front and 2, back view of leaf; 3, flower and bract; 4, flower with two calyx-lobes and petals removed, showing the base of two staminal bundles, and the style; 5, staminal bundle:—*all magnified.*

Tab. 6132.

IRIS LÆVIGATA.

Native of Japan and N. Eastern Asia.

Nat. Ord. IRIDACEÆ.—Tribe IRIDEÆ.

Genus IRIS, *Linn.; (Endl. Gen. Plant.*, p. 266).

IRIS *lævigata;* caule elato obtuse angulato foliato, foliis ½–⅔ poll. latis anguste lineari ensiformibus acuminatis utrinque viridibus costa prominula, scapo 1–2 flore, spathis 2–3-valvibus, valvis inæqualibus herbaceis elongato-lanceolatis acuminatis, floribus breviter pedicellatis maximis, pedicello ovario longiore, perianthii tubo crassiusculo segmentis exterioribus recurvis magnis late elliptico-ovatis obtusis ecristatis purpureis plaga basi aurea, interioribus parvis erectis conniventibus oblongis acutis, stigmatibus recurvis lineari-oblongis apice 2-lobis et dentatis.

IRIS lævigata, *Fisch., ex Turcz. Cat. Baikal,* No. 1119; *Fisch. et Mey Ind. Sem. Hort. Petrop.; Ledeb. Fl. Ross.,* vol. iv. p. 97; *Klatt in Linnœa,* vol. xxxiv. p. 616; *Maxim. Prim. Fl. Amurr.,* p. 271; *A. Gray Bot. Jap.,* p. 412; *Miquel Prol. Fl. Jap.,* 306.

I. Gemelini, *Ledeb. comment in Gmel. Fl. Sib. in Deuttscher, Bot. Ges. Regensb.,* vol. iii. p. 48.

I. Kæmpferi, *Sieb. ex Lemaire Ill. Hort.* t. 157.

I. versicolor, *Thunb. Fl. Jap.,* 34 (*ex Miquel, l.c.*) non *Linn.*

Whether under the indigenous form here figured, or the curious garden form called *I. Kæmpferi* var. *E. G. Henderson* (*Gard. Chron.*, 1874, p. 45), this beautiful hardy plant is likely to become as great a favourite in England as it is said to be in Japan. It was originally introduced by Von Siebold from Japan, and flowered in Verschaffelt's establishment at Ghent in 1857, when a very pale variety of it was figured by Lemaire in the "Illustration Horticole." As it there appeared under the name of *I. Kæmpferi* of Siebold, I suppose that this latter author identified it with the Sziti or Itz falz of Kæmpfer (*Amœn. Exot.* p. 873), a plant which Kæmpfer describes as an Iris with large double flowers, and which flowers during many days. Hasskarl (*Miquel Protus.* p. 306) says that it is the Itsi Katsi of the Japanese. Whatever may be its Japanese name or the history of that of Kæmpferi, it was no doubt first long previously described from Eastern Asiatic specimens by Fischer as *I. lævigata*. It is a native of East Siberia from the Baikal and Dahuria to Kamtschatka, the Amur district, and Korea, and it thence extends to the northern parts of Japan.

Mr. E. G. Henderson's variety, which I hope to figure soon, is a most remarkable and beautiful plant; it is a monstrous state, with six or more equal or unequal spreading perianth segments; for a description of which I must refer to Dr. Masters's article in the *Gardeners' Chronicle*, referred to above.

The specimen here figured was flowered in May last by G. Maw, F.L.S., in his rich garden at Benthall Hall, Shropshire, from roots received from Max Leichtlin of Baden Baden; it has also flowered at Kew for several years past.

DESCR. *Stem* two to three feet high, slender, obscurely angled. *Leaves* as long, one-half to two-thirds of an inch broad, narrow and slender, acuminate. *Scape* 1–3-flowered. *Spathes* two to four inches long, narrowly lanceolate, herbaceous, the outer shorter. *Flowers* three to five inches in diameter, varying from pale to deep red-purple; shortly pedicelled; pedicel shorter than the subterete ovary. *Perianth-tube* about three-quarters of an inch long; outer segments shortly clawed, broadly ovate-oblong, obtuse, reflexed, not crested, with a bright 3-cuspidate orange spot at the base of the limb; inner segments one to one and a half inches long, of the same purple colour, erect, conniving, sub-acute, oblong-lanceolate. *Stigmas* spreading, linear-oblong, with bifid incurved lobes.—*J. D. H.*

TAB. 6133.

POLYGONATUM VULGARE *var*. MACRANTHUM.

Native of Japan.

Nat. Ord. SMILACEÆ.—Tribe CONVALLARIEÆ.

Genus POLYGONATUM, *Tournef.* ; (*Endl. Gen. Pl.*, p. 154).

POLYGONATUM *vulgare;* caule arcuato acuto angulato, foliis alternis breviter petiolatis late ellipticis obtusis v. obtuse acuminatis subtus glaucescentibus 6-nerviis, floribus 1–4, perianthii supra medium subinflati lobis brevibus late orbiculato-ovatis viridibus apice obtuse apiculatis et incrassatis, filamentis glabris.

POLYGONATUM vulgare, *Desf.*; *A. Gray Bot. Jap.*, p. 413.

P. officinale, *All.*; *Maxim. Prim. Fl. Amur.*, 274; *Miquel Prol. Fl. Jap.*, 148.

CONVALLARIA Polygonatum, *Linn.*; *Thunb. Fl. Jap.*, p. 148.

VAR. *macranthum*, floribus 1¼ pollicaribus.

I have retained this plant as a variety of *P. vulgare* with much hesitation, doubting its proving even as a variety distinct from some already described forms of that variable plant. It is certainly not the same as Morren and Decaisne's *P. japonicum*, which they describe as having short solitary flowers with a campanulate perianth, and which is undoubtedly another form of *P. vulgatum*, a plant that extends from Western Europe (Norway and Spain) to the Western Himalaya, North East Asia and Japan, and which I suspect exists in Eastern N. America under one or more forms.

The size of the flower is perhaps the most noticeable feature of the plant here figured, though in that it is rivalled by both European and North Asiatic specimens; the inflation of the corolla above its middle and its slight contraction at the throat are other characters, which however disappear as the corolla withers and its lobes connive. Decaisne and Morren observe that the style exceeds the stamens in their *P. japonicum*, which is no doubt a sexual difference. In the form of its foliage it agrees best with the N. America *P. commutatum*, Dietr. and Otto, which has a terete stem. Lastly, having

NOVEMBER 1ST, 1874.

regard to the variability of the alternate-leaved *Polygonatums*, it would not surprise me to find that all were referable to two, the *P. vulgare* with a grooved stem, and *P. multiflorum* with terete stem.

The subject of the present plate has long been cultivated at Kew under the name of *P. japonicum*, and it flowers in April.

DESCR. *Rhizomes* stout, creeping. *Stems* one to one and a half feet high; stout, flexuous, green, angled and channelled. *Leaves* two to three inches long, alternate, subsessile, broadly elliptic, obtuse or obtusely acuminate, light green above, glaucous beneath, 7-nerved, quite glabrous. *Flowers* one to four; peduncles and pedicels longer or shorter than the perianth, very slender. *Perianth* one and a quarter inches long, terete, tubular, white, inflated slightly beyond the middle, contracted obscurely at the throat; lobes almost orbicular, with obtuse callous points, spreading, green. *Filaments* almost as long as the linear anthers, glabrous. *Ovary* globose; style filiform.—*J. D. H.*

Fig. 1, Flower laid open:—*somewhat magnified.*

Tab. 6134.

BLUMENBACHIA (Caiophora) contorta.

Native of Peru.

Nat. Ord. Loaseæ.

Genus Blumenbachia, *Schrad.*; (*Benth. & Hook. f. Gen. Pl.*, vol. i. p. 805).

Blumenbachia (Caiophora) *contorta;* caule volubili pilis urentibus patulis reflexisque hispido, foliis breviter petiolatis triangulari-v.-oblongo-ovatis pinnatifidis laciniis acutis acute inciso-dentatis lobatisve utrinque hispidis, pedunculis elongatis axillaribus, calycis lobis pinnatifido-lobatis, petalis patentibus, squamis cucullatis, staminodiis falcatis unidentatis, capsula ellipsoideo-oblonga 1½ pollicari.

Loasa contorta, *Lamk. Dict.*, vol. iii. p. 579; *Ill.* t. 426, f. 2; *DC. Prodr.*, vol. iii. p. 340; *Juss. in Ann. Mus.*, vol. v. p. 25, t. 3, f. 1; *Tratt. Archiv*, vol. i. p. 17, t. 33.

Caiophora contorta, *Presl Reliq. Haenk.*, vol. ii. p. 42; *Walp. Rep.*, vol. ii. p. 227, and vol. v. p. 781.

Although described by Jussieu and (copying him) by Trattinick as having a capsule a foot and a half long, "sesquipedalis," I have no doubt but that this is Lamarck's *Loasa contorta*, which that author describes as having a capsule about two inches long, as indeed it is figured by Jussieu. Lamarck's figure, again, a very indifferent one, represents the calyx-lobes as entire, though that author describes them as toothed. Presl, who founded the genus *Caiophora* on De Candolle's first section of *Loasa*, proposes besides the *C. contorta*, two other species from the Andes of Peru, *C. cirsiifolia* and *carduifolia;* but judging from his description and the figures he gives of *C. cirsiifolia*, I suspect that they are varieties of *C. contorta*, which, according to numerous specimens preserved in the Kew Herbarium, varies extremely in the breadth and amount of division of the leaves.

B. contorta is a native of Peru and Equador, where it ascends to an elevation of 12,000 feet; should it prove as hardy as the charming *B. lateritia* (*Loasa lateritia,* Tab. nost. 3632), it will be a very ornamental wall-plant in most parts

November 1st, 1874.

of England. It is probably, like that plant, a biennial. It was raised from Peruvian seeds by Messrs. Veitch, and flowered in their grounds in July of the present year.

DESCR. A climber, several feet high, hispid with spreading and recurved stinging bristles and shorter spreading hairs. *Leaves* shortly petioled, four to six inches long, triangular-oblong or -ovate, pinnatifid to the middle or to near the base, hispid, dark green; lobes few or many, broad or narrow, pinnatifidly lobed or toothed; pale blue-green beneath. *Peduncles* axillary, as long as or longer than the leaves, stout, hispid. *Flowers* one and a half to two inches in diameter. *Calyx-tube* short, obscure; lobes one-third to half an inch long, linear-oblong, pinnatifidly toothed or lobed. *Petals* bright brick-red, three-quarters of an inch long, spreading, obtuse, with a few bristles on the back. *Scales* a quarter of an inch long, cup-shaped, green, 3-toothed at the rounded tip, pubescent. *Staminodes* falcate, with one tooth on the margin, beyond which they are much narrowed and subulate. *Staminal-bundles* slender. *Capsule* (from Jussieu's figure) narrowly ellipsoid, an inch and a half long.—*J. D. H.*

Fig. 1, Scale and staminodes:—*magnified*.

Now Ready, Part IX., with 4 Coloured Plates, Royal 4to, price 5s.

ORCHIDS,
AND

How to Grow them in India & other Tropical Climates.

BY

SAMUEL JENNINGS, F.L.S., F.R.H.S.

Late Vice-President of the Agri-Horticultural Society of India.

NOW READY, Part II., 10s. 6d.

FLORA OF INDIA.

BY

DR. HOOKER, C.B., F.R.S.

Assisted by various Botanists.

NOW READY, Vol. VI., 20s.

FLORA AUSTRALIENSIS.

A Description of the Plants of the Australian Territory. By GEORGE BENTHAM, F.R.S., assisted by BARON FERDINAND MUELLER, C.M.G., F.R.S. Vol. VI. Thymeleæ to Dioscorideæ.

NOW READY.

LAHORE TO YARKAND.

Incidents of the Route and Natural History of the Countries traversed by the Expedition of 1870, under T. D. FORSYTH, Esq., C.B. By GEORGE HENDERSON, M.D., F.L.S., F.R.G.S., Medical Officer of the Expedition, and ALLAN O. HUME, Esq., C.B., F.Z.S., Secretary to the Government of India. With 32 Coloured Plates of Birds and 6 of Plants, 26 Photographic Views of the Country, a Map of the Route, and Woodcuts. Price 42s.

In the Press and shortly to be published, in one large Volume, Royal 8vo, with numerous Coloured Plates of Natural History, Views, Map and Sections. Price 42s.

To Subscribers forwarding their Names to the Publishers before publication, 36s.

ST. HELENA:

A

Physical, Historical, and Topographical Description of the Island,

INCLUDING ITS

GEOLOGY, FAUNA, FLORA, AND METEOROLOGY.

BY

JOHN CHARLES MELLISS, C.E., F.G.S., F.L.S.

LATE COMMISSIONER OF CROWN PROPERTY, SURVEYOR AND ENGINEER OF THE COLONY.

L. REEVE & Co., 5, Henrietta Street, Covent Garden.

DEDICATED BY SPECIAL PERMISSION TO H.R.H. THE
PRINCESS OF WALES.

NOW READY, Complete in Six Parts, 21s. each, or in One Vol., imperial folio, with 30 elaborately Coloured Plates, forming one of the most beautiful Floral Works ever published, half morocco, gilt edges, £7 7s.

A MONOGRAPH OF ODONTOGLOSSUM.

A Genus of the Vandeous section of Orchidaceous Plants. By JAMES BATEMAN, F.R.S., F.L.S., Author of "The Orchidaceæ of Mexico and Guatemala."

L. REEVE & Co., 5, Henrietta Street, Covent Garden.

NOW READY, Vol. 3, with 551 Wood Engravings, 25s.

THE NATURAL HISTORY OF PLANTS.

By Prof. H. BAILLON, P.L.S., Paris. Translated by MARCUS M. HARTOG, B.Sc., Lond., B.A., Scholar of Trinity College, Cambridge. Contents:—Menispermaceæ, Berberidaceæ, Nymphæaceæ, Papaveraceæ, Capparidaceæ, Cruciferæ, Resedaceæ, Crassulaceæ, Saxifragaceæ, Piperaceæ, Urticaceæ.

L. REEVE & Co., 5, Henrietta Street, Covent Garden.

QUADRANT HOUSE,

74, REGENT STREET, AND 7 & 9, AIR STREET, LONDON, W.

AUGUSTUS AHLBORN,

Begs to inform the Nobility and Gentry that he receives from Paris, twice a week, all the greatest novelties and specialties in Silks, Satins, Velvets, Shawls, &c., and Costumes for morning and evening wear. Also at his establishment can be seen a charming assortment of robes for Brides and Bridesmaids, which, when selected, can be made up in a few hours. Ladies will be highly gratified by inspecting the new fashions of Quadrant House.

From the *Court Journal*:—"Few dresses could compare with the one worn by the Marchioness of Bute at the State Concert at Buckingham Palace. It attracted universal attention, both by the beauty of its texture, and the exquisite taste with which it was designed. The dress consisted of a rich black silk tulle, on which were artistically embroidered groups of wild flowers, forming a most elegant toilette. The taste of the design, and the success with which it was carried out, are to be attributed to the originality and skill of Mr. AUGUSTUS AHLBORN."

Third Series.

No. 360.

VOL. XXX. DECEMBER. [*Price 3s. 6d. col^d 2s. 6d. plain.*

OR No. 1054 OF THE ENTIRE WORK.

CURTIS'S
BOTANICAL MAGAZINE,

COMPRISING

THE PLANTS OF THE ROYAL GARDENS OF KEW,

AND OF OTHER BOTANICAL ESTABLISHMENTS IN GREAT BRITAIN,
WITH SUITABLE DESCRIPTIONS;

BY

JOSEPH DALTON HOOKER, M.D., C.B., F.R.S., L.S., &c.

Director of the Royal Botanic Gardens of Kew.

Nature and Art to adorn the page combine,
And flowers exotic grace our northern clime.

LONDON:
L. REEVE & CO., 5, HENRIETTA STREET, COVENT GARDEN.
1874.

[*All rights reserved.*]

ROYAL HORTICULTURAL SOCIETY.

MEETINGS AND SHOWS IN 1874.

December 2. (Fruit and Floral Meeting.)

RE-ISSUE of the THIRD SERIES of the BOTANICAL MAGAZINE.

Now ready, Vols. I. to XII., price 42s. each (to Subscribers for the entire Series 36s. each).

THE BOTANICAL MAGAZINE, Third Series. By Sir WILLIAM and DR. HOOKER. To be continued monthly.

Subscribers' names received by the Publishers, either for the Monthly Volume or for Sets to be delivered complete at 36s. per Volume, as soon as ready.

L. REEVE & Co., 5, Henrietta Street, Covent Garden.

BOTANICAL PLATES;
OR,
PLANT PORTRAITS.

IN GREAT VARIETY, BEAUTIFULLY COLOURED, 6d. and 1s. EACH.

List of 2000 Species, one stamp.

L. REEVE & Co., 5, Henrietta Street, Covent Garden.

THE FLORAL MAGAZINE.

NEW SERIES, ENLARGED TO ROYAL QUARTO.

Figures and Descriptions of the Choicest New Flowers for the Garden, Stove, or Conservatory. Monthly, with 4 Coloured Plates, 3s. 6d. Annual Subscription, 42s.

L. REEVE & Co., 5, Henrietta Street, Covent Garden.

FLORAL PLATES,

BEAUTIFULLY COLOURED, 6d. AND 1s. EACH.
New Lists of 600 Varieties, one stamp.

L. REEVE & Co., 5, Henrietta Street, Covent Garden.

Tab. 6135.

RHEUM OFFICINALE.

Native of Eastern Tibet and Western China.

Nat. Ord. POLYGONACEÆ.—Tribe PTERYGOCARPÆ.

Genus RHEUM, *Linn.*; (*Meissn.*, *in DC. Prodr.*, v. xiv. p. 32).

RHEUM *officinale;* caule brevi robusto diviso apice folioso, foliis amplis orbiculari-ovatis cordatisve villosulis subpalmatim breviter 3-7-lobatis, lobis incisis lobulisque acutis, ochrea obovoidea densum fissa, petiolo robusto pubescente intus haud sulcato, ramis floriferis foliosis paniculatim ramulosis, paniculæ erectæ ramulis patenti-recurvis, ultimis floriferis spiciformibus nutantibus densifloris, floribus breviter gracile pedicellatis, pedicello basin versus articulato, perianthii foliolis late oblongis apice rotundatis interioribus paulo majoribus, staminibus 9 inclusis, disco annulari crenulato, stigmatibus orbiculatis, achænio late oblongo alis membranaceis obscure crenulatis nucleo duplo longioribus et laterioribus.

RHEUM officinale, *Baillon in Mém. de l'Association Française pour l'Avancement des Sciences*, Sept. 1, Bordeaux, 1872, p. 514, t. x. (Translated in *Trimen Journ. Bot.*, 1872, p. 379); *Adansonia*, vol. x. p. 246; *Carrière, Rev. Hortic.*, 1874, p. 93; *Fluck. & Hanbury, Pharmacog.*, p. 442; *Gard. Chron.*, 1874, v. 1. p. 346.

According to the evidence hitherto obtained, this grand plant (which is certainly the handsomest of all the *Rheums*, except the Himalayan *R. nobile*) is that which produces much, if not all the Turkey Rhubarb of the pharmacopœia. It is a native of and also cultivated in Eastern and South-eastern Tibet, and was sent thence by the French missionaries to M. Dabry, the French Consul at Hankow. M. Dabry sent plants to M. Soubeiran, Secretary of the Jardin d'Acclimatisation of Paris, where they flowered at Montmorency in 1871.

An excellent history of this plant is given in Fluckiger and Hanbury's "Pharmacographia," quoted above, from which it appears not to be certain that the true Turkey rhubarb of commerce is derived exclusively from this plant, though the evidence of the missionaries who discovered it, that it is the main source of that drug, is supported by the fact that there is no important discrepancy between this *R. officinale* and the imperfect and scanty accounts and figures of the Chinese authors and early French missionaries. From the same work we learn that the drug was known to the Chinese long anterior to the Christian era, and was described in a work dedicated to the Emperor Shen-nung, the father of Chinese agriculture and medicine, who lived about 2700 B.C. Also that Marco Polo is the only traveller who has visited the districts yielding rhubarb, in the mountains of one of which (Tangut) he describes it as growing in great abundance; this, however, is an error, for an account of it will be found in the Travels of Bell of Antermony (vol. i. p. 384—387), who found it in Mongolia, growing abundantly near marmot burrows. One of its most remarkable characteristics is its stout very distinct stem, which, and not the root, is considered to be the source of the rhubarb in the view of M. Baillon, and no doubt correctly.

The rhubarb plant inhabits a vast area of Eastern Tibet and Western China, abounding in high plateaus, especially in spots enriched by old encampments. The plant here figured was sent to the Royal Gardens by M. Soubeiran, and flowered in June last, both at Kew and at Mr. Hanbury's garden at Clapham.

DESCR. *Stem* as thick as the arm, four to ten inches high, divided into several leafing and flowering crowns. *Leaves* one to three feet in diameter, orbicular-ovate or cordate, shallowly 3- to 7-lobed, pubescent or subvillous, lobes acute and acutely irregularly toothed; nerves stout beneath, flabellate; petiole nearly terete; ochrea split. *Flowering-stems* two to five feet high, erect, stout, leafy, pubescent, bright-green, paniculately branched; branches spreading; flowering branchlets spreading and drooping, three to five inches long, spiciform, very densely clothed with flowers. *Flowers* one-quarter inch in diameter, green; pedicels slender, fascicled, jointed near the base. *Perianth-segments* broadly oblong, rounded at the tips, the inner larger, erect at the edges. *Stamens* nine, as long as the perianth, hypogynous. *Disk* thick, annular, obscurely 3-lobed, crenulate. *Stigmas* orbicular, peltate. *Fruit* half an inch long, broadly oblong, emarginate, bright red; wings longer and broader than the nucleus.—*J. D. H.*

Fig. 1, Reduced view of whole plant; 2, portion of leaf, and 3, of flowering branch, both of the natural size; 4, flower; 5, stamen and pistil; 6, pistil and disk:—*all magnified*; 7, fruiting branchlet, not seen; 8, fruit:—*magnified*.

TAB. 6136.

EPISCIA FULGIDA.

Native of New Grenada.

Nat. Ord. GESNERIACEÆ.—Tribe BESLERIEÆ.

Genus EPISCIA, *Mart.*; (*DC. Prodr.*, vol. vii. p. 546).

EPISCIA *fulgida*; repens, stolonifera, tota pilis flaccidis villose, foliis ellipticis elliptico-ovatisve subacutis basi rotundatis v. cordatis crenulato-serratis convexis superne bullatim reticulatis, petiolo brevi crassiusculo, floribus solitariis, pedunculo petiolo longiore, sepalis $\frac{1}{2}$–$\frac{3}{4}$-pollicaribus spathulato-oblongis subserratis, corollæ læte lateritiæ tubo 1$\frac{1}{2}$-pollicari hirsuto, limbo lobis rotundatis erosis, staminibus inclusis, ovario hirsuto.

CYRTODEIRA fulgida, *Lind. Cat.* No. 90, p. 5; *et in L'Illustr. Hortic.*, t. 131.

I have little doubt as to this being the plant described by Linden as *Cyrtodeira fulgida,* of which the figure in the "Illustration Horticole" is excellent, though in the description the blade of the leaf is described as scarcely longer than the petiole, and although the leaves want the pale band along the midrib and principal nerves, which render Linden's form of it so valuable for decorative purposes. It is a very close ally of the Brazilian *E. reptans* (Mart. Nov. Gen., t. 217); but *E. fulgida* is a much larger plant, and has shorter petioles, and differently shaped sepals, which are not entire; it also comes from a very different country. Another congener is the *Achimenes cupreata* Hook. (Tab. nost. 4312), upon which Hanstein (L. v. xxvi. p. 206) founded his genus *Cyrtodeira,* distinguishing it from *Episcia* by the form and lesser curvature of the corolla-tube, a character that does not hold in the various species. *Episcia fulgida* is a native of New Grenada, whence it was first introduced by M. Linden. I am indebted to Mr. Williams for the specimen here figured, which flowered in his establishment in July last.

DESCR. *Stem* creeping, branched, stoloniferous, as thick as a goose-quill, and, as well as the whole plant, clothed with a soft villous pubescence. *Leaves* three to five inches long,

DECEMBER 1ST, 1874.

elliptic or elliptic-ovate, acute, crenulate-serrate, convex and bullately reticulated on the upper surface, dark emerald-green, paler along the midrib, inclined to coppery, especially the young ones; petiole stout, about one-eighth the length of the blade. *Peduncles* axillary, solitary, stout, one to two inches long. *Calyx* gibbous at the base, one-half to three-quarters inch long, campanulate; sepals spathulate-oblong, rounded and crenate towards the recurved tips. *Corolla* bright and almost vermilion-red; tube hirsute, one and a-half inch long, cylindric, nearly straight; limb one inch in diameter, nearly equal; lobes rounded, irregularly toothed, pubescent towards the throat. *Stamens* included, filaments very slender; anthers small, adnate to the large connective. *Ovary* broadly ovoid, hirsute; gland emarginate.—*J. D. H.*

Fig. 1, Cape; 2, corolla laid open; 3, top of stamen; 4 and 5, front and side view of ovary:—*all magnified.*

Tab. 6137.

BOUCEROSIA MAROCCANA.

Native of Marocco.

Nat. Ord. ASCLEPIADEÆ.—Tribe STAPELIEÆ.

Genus BOUCEROSIA, *Wight & Arn.*; (*Benth. & Hook. f. Gen. Plant.*, vol. ii. ined.).

BOUCEROSIA *maroccana*; ramis tetragonis marginibus angulato sinuatis, foliis trulliformibus acutis, corollæ lobis nudis tubo intus hirsuto, coronæ stamineæ processubus 5 incurvis gynostegio incumbentibus cum 10 erectis geminatim collateralibus capitatis alternantibus.

A near ally of *B. Gussoniana* (*Apteranthes Gussoniana*, Tab. nost. 5087), of Algeria, and so like it that I long hesitated before deciding upon figuring and describing it as new; besides, however, the differences of habit, which are more easily seen than described, there are so many other differential characters that, taken altogether, it is impossible to unite this with the Algerian plant. The angles of the stem, instead of being faintly undulate, are longitudinally divided into broadly triangular lobes, each tipped with a minute leaf, which instead of being sessile adnate and upcurved, is trowel-shaped, contracted at the base, and usually deflexed. The flowers are fewer, on longer pedicels; the corolla-lobes want the long cilia, being quite naked, and are shorter, not so reflexed, and of a clearer purple, with fewer yellow bars, and the base of the tube inside is densely velvety. But the greatest difference is in the staminal crown, which in *B. Gussoniana* presents five capitate incurved processes, each with a knob on each side at the base; but in this the five incurved processes are inflexed and incumbent upon the stigma, whilst the lateral knobs are elevated on erect stalks. How far any or all these characters are variable can only be known by a long and careful study. The probability of their proving constant is rendered more probable by the wide difference of habit of the two plants, the *A. Gussoniana* being a native of saline situations in Sicily, Spain, and the Algerine coast; whilst *A. maroccana* inhabits the much lower latitudes of Mogado in Marocco. Here it was found on the rocky islet of Mogador by Messrs. Maw, Ball, and myself, and also elsewhere along the coast near the town, and introduced into Kew, where it flowered in July.

The genus *Apteranthes* is merged in *Boucerosia* (itself possibly referrible to *Piaranthus*) by Bentham in the forthcoming volume of the Genera Plantarum.

DESCR. *Branches* prostrate, six to ten inches long, by about one broad, 4-sided, the sides deeply sunk, the angles cut into broad subtriangular lobes, with an acute sinus, one-half to three-quarter inch long. *Leaves* on the angular summits of the lobes of the stem-angles, one-tenth inch long, trowel-shaped, contracted at the base, ciliolate. *Flowers* two to six in an umbel, pedicel one-quarter inch long, and subulate calyx-teeth green. *Corolla* one-half to two-thirds inch diameter, spreading, 5-lobed to about the middle; lobes triangular, subacute, quite glabrous, pale green outside, dark red-purple within streaked transversely towards the base and around the cup with yellow; tube densely villous. *Crown* of 5 inflexed processes that cover the staminal crown, and 10 erect capitate processes in pairs between the inflexed ones.—J. D. H.

Fig. 1, Leaf; 2, outside, and 3, inside view of corolla; 4, gynostegium; 5, pollen-masses:—*all magnified.*

TAB. 6138.

ONCIDIUM ZEBRINUM.

Native of Venezuela.

Nat. Ord. ORCHIDEÆ.—Tribe VANDEÆ.

Genus ONCIDIUM, *Swartz;* (*Lindl. Fol. Orchid.*, Oncidium).

ONCIDIUM (Cyrtochilum) *zebrinum;* rhizomate robusto repente, pseudobulbis compressis 4–5-pollicaribus ovato-lanceolatis lævibus, foliis 6–9-pollicaribus ensiformi-lanceolatis acuminatis carinatis nervosis, panicula longissima robusta flexuosa, bracteis spathaceis ovato-oblongis obtusis, perianthii foliolis albis rubro-fasciatis, petalis sepalisque conformibus ligulato-oblanceolatis crispato-undulatis, labello parvo e basi dilatato carunculato in laminam recurvam ovatam angustato, columna brevi recurva antice tumida sulcata utrinque apicem versus unidentata.

O. zebrinum, *Rchb. f. in Seem. Bompland,* 1854; *Lindl. Fol. Orchid. Oncid.* No. 16; *Rchb. f. in Gard. Chron.,* 1872, p. 1355.

ODONTOGLOSSUM zebrinum, *Rchb. f. in Linnæa,* vol. xxii. p. 849; *Lindl. Fol. Orchid.* Odontoglossum, No. 40.

A very attractive plant from the pure white of the perianth, with its red-purple bars, and the fine gamboge-yellow of the bars of the lip; at least such are the attractions of the variety figured here. But Reichenbach describes a form in which the whole disk of the sepals is violet, and with only one violet spot at the base of each petal. In the length of the panicle it exceeds all other species I have seen growing; in the specimen here figured it was twelve feet long. Mr. Burbridge, who communicated the plant to me from the garden of Sir William Marriott, of Dover House, Blandford, observes, that the pseudo-bulbs are quite like those of *O. macranthum* (Tab. nost. 5743).

Oncidium zebrinum has been sent home, living or dried, by various collectors, and was first flowered, according to Professor Reichenbach, by Mr. Bull, in 1872.

DESCR. *Rhizome* stout, creeping, as thick as a goose-quill, with lanceolate brown scarious sheaths about one inch long. *Pseudo-bulbs* three and a-half to four inches long, by one and a-half to one and three-quarter inch broad, narrow ovoid, compressed, green, grooved when old or dry. *Leaves* six to nine inches long, between ensiform and lanceolate, acuminate, striated, keeled, deep green, paler beneath. *Panicle* sometimes twelve feet long, peduncle and rachis green, terete, as thick as a crow-quill and upwards, very flexuous but rigid, branches six inches long. *Bracts* one-half to three-quarter inch long, oblong-lanceolate, brown, dry. *Pedicel* together with the slender ovary one inch long. *Perianth* one and a-half to one and three-quarter inch in diameter. *Sepals* and *Petals* very similar, narrowly obovate or oblanceolate, or somewhat spathulate, waved and crisped, white with violet-red bars from the base to the middle. *Lip* much smaller than the petals, base broad subquadrate, irregularly thickened and warted, the centre yellow, the edges barred; from this base the lip suddenly contracts into a triangular thick reflexed somewhat concave limb, which is white speckled with dull-red. *Column* short, tumid, and grooved in part, with a horizontal process on each side at the tip.—*J. D. H.*

Fig. 1, View of ovary, column, and lip:—*magnified.*

TAB. 6139.

FUCHSIA PROCUMBENS.

Native of New Zealand.

Nat. Ord. ONAGRARIEÆ.

Genus FUCHSIA, *Linn.*; (*Benth. & Hook. f. Gen. Plant.*, vol. 1, p. 790).

FUCHSIA *procumbens;* glaberrima, prostata, caulibus filiformibus elongatis, foliis longe graciliter petiolatis ovatis v. orbiculato v. cordato-ovatis obtusis obscure sinuato-dentatis inter nervos non reticulatis, floribus solitariis axillaribus erectis, calycis tubo cylindraceo-campanulato basi rotundato non inflato, lobis oblongis obtusis, petalis 0.

FUCHSIA procumbens, *R. Cunn. in A. Cunn. Bot. Fl. Nov. Zel. in Ann. Nat. Hist.*, vol. iii. p. 31; *Hook. Ic. Pl.* t. 421; *Hook. f. Fl. Nov. Zeld.*, vol. i. p. 57; *Handb. New Zeald. Flor.*, p. 76 and 728; *Masters in Gard. Chron.*, Sept. 1874. (Planta ♀).

F. Kirkii, *Hook. f. in Ic. Plant.*, t. 1083 (Planta ♂).

This curious little plant, so unlike a *Fuchsia* in habit and colour of the flower, was discovered in 1834 by Richard Cunningham in the northern Island of New Zealand, on the shores of the east coast, opposite the Cavalhos Islands, growing on the sandy beach, where it has since been gathered by Colenso. It has also been found on the Great Barrier Islands by Mr. Kirk in two localities, both near the sea. This latter I distinguished as *F. Kirkii*, relying on the length of the style and large capitate stigma, which I now find evidently a sexual character. I also attributed to the true *F. procumbens* lanceolate sepals; in this latter character I now find I was deceived by the Herbarium containing a mixture of specimens of what I take to be a slender variety of *F. Colensoi* (Handbook of New Zealand Flora, p. 728) with those of *F. procumbens.* Like all New Zealand plants, the *Fuchsias* are extraordinarily variable, and the two small species (*procumbens* and *Colensoi*) are certainly bisexual. Of the three New Zealand species, *F. excorticata* and *Colensoi* have the leaves reticulated between the nerves; the corolla-tube, inflated at the base, then suddenly contracted and dilated

DECEMBER 1ST, 1874.

again into a funnel-shaped limb, with lanceolate spreading lobes, and have minute petals; whereas *F. procumbens* has leaves without reticulation, has a cylindric corolla-tube with linear-oblong lobes which are sharply reflexed on the tube; it has also often shorter petioles than *F. Colensoi*, and is apetalous. *F. Colensoi* may be divisible into two species; one more robust with the calyx three-quarters of an inch long, the other as slender and trailing as *F. procumbens*, with the calyx one-third to one-half inch long; but this can only be determined by studying the plant in all its states: the larger form is possibly only a small state of *F. excorticata*.

F. procumbens was introduced into England by the late Mr. Williams, of Hendon, many years ago; and again by Mr. J. Blackett, of Egham, from whose plant Mr. Burbridge sent me an excellent drawing in August last, in which, however, the flowers are represented as pendulous. About the same time Mr. Kinghorn, of Richmond, brought me a beautiful plant of it, from which the accompanying drawing was made by my very accomplished friend, J. F. Moggridge, F.L.S., and which is here reproduced by Mr. Fitch.

DESCR. *Stems* filiform, trailing, often many feet long. *Leaves* one-half to three-quarters of an inch long, ovate or cordate, rarely orbicular, obscurely sinuate-toothed, membranous, pale green above, almost white beneath, not reticulated between the principal nerves; petiole filiform, longer than the blade. *Flowers* solitary, axillary, erect, pedicels one-fourth to one-half inch long. *Calyx-limb* longer than the pedicel, cylindric, rounded but not inflated at the base, pale orange-yellow; lobes spreading, then reflexed on the tube, linear-oblong, obtuse, dark purple, green at the base. *Stamens* in the male plant on slender filaments; anthers blue; those of the female with short filaments. *Ovary* ovoid; style in the male plant included with a small stigma, in the female exserted with a large capitate stigma.—*J. D. H.*

Fig. 1, Flower, with corolla laid open:—*magnified*.

INDEX

To Vol. XXX. of the THIRD SERIES, or Vol. C. of the Work.

Pl.
- 6117 Achillea ageratifolia.
- 6092 Aconitum heterophyllum.
- 6087 Arabis blepharophylla.
- 6079 Bambusa striata.
- 6086 Bauhinia natalensis.
- 6091 Beschorneria Tonelii.
- 6134 Blumenbachia contorta.
- 6119 Bolbophyllum Dayanum.
- 6137 Boucerosia maroccana.
- 6114 Brachysema undulatum.
- 6123 Brodiæa volubilis.
- 6104 Calanthe curculigoides.
- 6111 Campsidium chilense.
- 6130 Cerinthe gymnandra.
- 6107 Chrysanthemum Catananche.
- 6120 Cinnamodendron corticosum.
- 6128 Citrus Aurantium, var. japonica.
- 6090 Colchicum Parkinsoni.
- 6078 Colchicum speciosum.
- 6113 Crinum Moorei.
- 6103 Crocus cancellatus.
- 6115 Decabelone elegans.
- 6121 Drosera Whittakerii.
- 6097 Echinocactus Cummingii.
- 6094 Epidendrum criniferum.
- 6098 Epidendrum Lindleyanum.
- 6136 Episcia fulgida.
- 6108 Erica Chamissonis.
- 6080 Fagræa zeylanica.
- 6139 Fuchsia procumbens.
- 6081 Gaillardia Amblyodon.
- 6105 Grevillea fasciculata.
- 6083 Iris Douglasiana.

Pl.
- 6132 Iris lævigata.
- 6110 Iris olbiensis.
- 6118 Iris tectorum.
- 6116 Kniphofia Rooperi.
- 6106 Lessertia perennans.
- 6126 Lilium maculatum.
- 6131 Melaleuca Wilsoni.
- 6077 Mesembryanthemum truncatellum.
- 6088 Nunnezharia geonomæformis.
- 6085 Odontoglossum Roezlii.
- 6084 Odontoglossum roseum.
- 6138 Oncidium zebrinum.
- 6093 Panax sambucifolius.
- 6129 Passiflora manicata.
- 6122 Pentstemon humilis.
- 6125 Pogonia discolor.
- 6133 Polygonatum vulgare.
- 6112 Pyrus baccata.
- 6100 Regelia ciliata.
- 6135 Rheum officinale.
- 6089 Rhipsalis Houlletii.
- 6095 Rhopala Pohlii.
- 6109 Romanzoffia sitchensis.
- 6102 Saxifraga florulenta.
- 6074 Saxifraga peltata.
- 6127 Scorzonera undulata.
- 6099 Senecio Anteuphorbium.
- 6101 Senecio Doronicum.
- 6082 Stapelia Corderoyi.
- 6076 Steudnera colocasiæfolia.
- 6124 Tacca artocarpifolia.
- 6075 Xanthorrhæa quadrangulata.
- 6096 Xiphion Sisyrinchium.

Now Ready, Part X., with 4 Coloured Plates, Royal 4to, price 5s.

ORCHIDS,
AND

How to Grow them in India & other Tropical Climates.
BY

SAMUEL JENNINGS, F.L.S., F.R.H.S.

Late Vice-President of the Agri-Horticultural Society of India.

NOW READY, Part II., 10s. 6d.

FLORA OF INDIA.
BY

DR. HOOKER, C.B., F.R.S.

Assisted by various Botanists.

NOW READY, Vol. VI., 20s.

FLORA AUSTRALIENSIS.

A Description of the Plants of the Australian Territory. By GEORGE BENTHAM, F.R.S., assisted by BARON FERDINAND MUELLER, C.M.G., F.R.S. Vol. VI. Thymeleæ to Dioscorideæ.

NOW READY.

LAHORE TO YARKAND.

Incidents of the Route and Natural History of the Countries traversed by the Expedition of 1870, under T. D. FORSYTH, Esq., C.B. By GEORGE HENDERSON, M.D., F.L.S., F.R.G.S., Medical Officer of the Expedition, and ALLAN O. HUME, Esq., C.B., F.Z.S., Secretary to the Government of India. With 32 Coloured Plates of Birds and 6 of Plants, 26 Photographic Views of the Country, a Map of the Route, and Woodcuts. Price 42s.

In the Press and shortly to be published, in one large Volume, Royal 8vo, with numerous Coloured Plates of Natural History, Views, Map and Sections. Price 42s.

To Subscribers forwarding their Names to the Publishers before publication, 36s.

ST. HELENA:
A

Physical, Historical, and Topographical Description of the Island,

INCLUDING ITS

GEOLOGY, FAUNA, FLORA, AND METEOROLOGY.

BY

JOHN CHARLES MELLISS, C.E., F.G.S., F.L.S.

LATE COMMISSIONER OF CROWN PROPERTY, SURVEYOR AND ENGINEER OF THE COLONY.

L. REEVE & CO., 5, Henrietta Street, Covent Garden.

DEDICATED BY SPECIAL PERMISSION TO H.R.H. THE
PRINCESS OF WALES.

NOW READY, Complete in Six Parts, 21s. each, or in One Vol., imperial folio, with 30 elaborately Coloured Plates, forming one of the most beautiful Floral Works ever published, half morocco, gilt edges, £7 7s.

A MONOGRAPH OF ODONTOGLOSSUM.

A Genus of the Vandeous section of Orchidaceous Plants. By JAMES BATEMAN, F.R.S., F.L.S., Author of "The Orchidaceæ of Mexico and Guatemala."

L. REEVE & Co., 5, Henrietta Street, Covent Garden.

NOW READY, Vol. 3, with 551 Wood Engravings, 25s.

THE NATURAL HISTORY OF PLANTS.

By Prof. H. BAILLON, P.L.S., Paris. Translated by MARCUS M. HARTOG, B.Sc., Lond., B.A., Scholar of Trinity College, Cambridge. Contents :—Menispermaceæ, Berberidaceæ, Nymphæaceæ, Papaveraceæ, Capparidaceæ, Cruciferæ, Resedaceæ, Crassulaceæ, Saxifragaceæ, Piperaceæ, Urticaceæ.

L. REEVE & Co., 5, Henrietta Street, Covent Garden.

QUADRANT HOUSE,

74, REGENT STREET, AND 7 & 9, AIR STREET, LONDON, W.

AUGUSTUS AHLBORN,

Begs to inform the Nobility and Gentry that he receives from Paris, twice a week, all the greatest novelties and specialties in Silks, Satins, Velvets, Shawls, &c., and Costumes for morning and evening wear. Also at his establishment can be seen a charming assortment of robes for Brides and Bridesmaids, which, when selected, can be made up in a few hours. Ladies will be highly gratified by inspecting the new fashions of Quadrant House.

From the *Court Journal*:—"Few dresses could compare with the one worn by the Marchioness of Bute at the State Concert at Buckingham Palace. It attracted universal attention, both by the beauty of its texture, and the exquisite taste with which it was designed. The dress consisted of a rich black silk tulle, on which were artistically embroidered groups of wild flowers, forming a most elegant toilette. The taste of the design, and the success with which it was carried out, are to be attributed to the originality and skill of Mr. AUGUSTUS AHLBORN."

Lightning Source UK Ltd.
Milton Keynes UK
UKHW022016230222
399153UK00003B/117